The Age of Learjet

Adventures, Innovations, and Sky-High Dreams

Peter G. Hamel and Gary D. Park

The Age of Learjet
Adventures, Innovations, and Sky-High Dreams.

Copyright © 2023 by Peter G. Hamel and Gary D. Park

Edited by Preston Park

Cover design by Preston Park

Published by Winning Growth Systems, LLC
Grapevine, TX, USA
in 2023.

ISBN: 979-8-9892160-0-0

Library of Congress Control Number: 2023919996

All rights reserved. No part of this book may be reproduced or transmitted in any form or by any means without prior written permission from the publisher.

Includes material reprinted from: *The Learjet History* (Springer, 2022), copyright © 2022 by Peter G. Hamel and Gary D. Park, used with permission from Springer Nature Switzerland AG. © Springer Nature Switzerland AG 2023.

This book is intended to offer general guidance. It is sold with the understanding that the publisher and author are not engaged in rendering legal, accounting, or other professional services. If legal advice or other expert assistance is required, the services of a competent professional should be sought.

Acknowledgements

Hello, readers! We want to give a big shout-out to all the awesome people who helped make this book possible. There were so many aviation enthusiasts who joined us on this journey, and we're super grateful!

We'd love to name every single person who helped us, but that's a pretty big list. We want to say a huge THANK YOU to everyone who supported us in big and small ways. You rock!

But there are a few people we really want to highlight:

- Lou Knotts from the Calspan Corporation let us use cool documents to talk about their amazing Learjet 24/25/31 fleet. These jets are one-of-a-kind and super high-tech!

- Dr. Ravindra Jategaonkar, a super smart scientist, looked through *The Learjet History* and gave us really smart feedback. His help made our book even better!

- Dr. Herman Rediess wrote a special message at the beginning of *The Learjet History*, saying how awesome we are. We're really thankful for his kind words!

- Brian Barents, who was the CEO of Learjet from 1980 to 1996. Brian took a peek at our chapter on the Learjet 31 and shared his stellar insights to make sure we got the facts right. We're totally grateful for his help!

- Some former Learjet employees like Al Higdon, Mike Abla, Stan Blankenship, and others helped us with their expert knowledge. They even reviewed our work to make sure everything was accurate and cool.

- Rick Durden shared his flying experiences with Learjets, including some safety tips. He writes cool stuff on the internet that pilots love to read!

- Dick Kovich provided invaluable proofreading of *The Learjet History* and corrected many important facts. His attentive review ensured accuracy throughout the book.

- Paul Bowen took amazing pictures of Learjets flying in the sky. His photos look awesome and he's a rockstar in the aviation world!

- Preston Park helped us with computer stuff. He's like a computer wizard!
- Georges Bridel, a cool Swiss engineer, told us some cool stuff about the Swiss combat aircraft P-16. It's always neat to learn about planes from around the world!

We also want to give a big thumbs-up to all the NASA centers that helped Learjet succeed. They're like the superheroes of aviation! Richard Travis Whitcomb designed special winglets for Learjets that helped them fly better and break records. Ames and Langley NASA centers were like Learjet's secret helpers! And to our amazing wives, Judy and Hanni, you're the best for understanding and supporting us while we worked on this book.

Last but not least, a huge thanks to Springer International Publisher for believing in the original book, *The Learjet History, Beginnings, Innovations, and Utilizations*, and helping us bring it to life. Leontina Di Cecco and Gowtham Chakravarthy and the Springer Nature Production Team made everything run smoothly.

Thanks to all of you for joining us on this journey through Learjet history. We hope you enjoy reading this book as much as we enjoyed making it!

With gratitude,

Peter and Gary

Preface

Greetings, fellow adventurers! We're Peter and Gary, your friendly guides through the thrilling story of the Learjet. We want to share an awesome story with you - the history of the Learjet!

Get ready to embark on an epic journey back in time, where innovation, heroes, and sky-high dreams take center stage.

Believe it or not, in 2022, the Learjet celebrated its 60th anniversary in Wichita, Kansas. To honor this remarkable milestone, Kevin Burns of the American Institute of Aeronautics and Astronautics (AIAA) sent an email about presenting a paper.

We dove deep into research of the history of Learjet and the daring pilots who soared through incredible Learjet missions. But guess what? The stories we uncovered from over 250 sources were so amazing that they couldn't be contained in a simple paper.

So, we did something extraordinary. We transformed our research into a book, just for you! Here, you'll find an abundance of captivating tales that unveil the mesmerizing history of Learjet, from its early days to the unforgettable impact it had on the world of aviation.

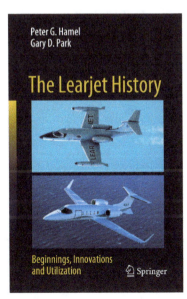

Hold on tight, because there's more! This book is a special version of a larger work we crafted, *The Learjet History: Beginnings, Innovations and Utilization*. We took all the thrilling parts, added a dash of adventure, and created these action-packed stories you're about to devour.

Discover captivating articles and find out more about our comprehensive book *The Learjet History* at https://learjethistory.com. We only noted a few of the major sources at the end of this book. So, if you want to see the full academic list and technical information, you can find them in the original work.

Strap yourselves in, readers, because we're about to take off on an exhilarating journey through time and aviation. Let's explore the tales of those who reached for the skies and shaped the world of Learjet.

We're thrilled to have you on board, and we hope you enjoy this adventure as much as we enjoyed crafting it.

Fasten your seatbelts and let's soar together,

Peter and Gary

Contents

Chapter 1 Introducing the Learjet Adventure ... 1
    Welcome! ... 1
    Inset: Discover Innovation ... 3
    Key Takeaways ... 4

Chapter 2 Sky-High Dreams ... 5
    Introduction ... 5
    The Need for Speed ... 5
    Meet the Mavericks ... 6
    A Brilliant New Partner ... 7
    Dr. Studer's Ingenious Pivot ... 9
    Let Your Mind Soar ... 10
    Bumpy Skies Ahead ... 11
    Kansas or Bust ... 13
    Wheels Up! ... 13
    Dreams Take Flight ... 14
    Inset: What is "Boundary Layer Control"? ... 15
    Key Takeaways ... 16

Chapter 3 Pushing the Limits ... 17
    Introduction ... 17
    Blazing a Trail with the Speedy 23 ... 18
    Tragedy Sparks Safer Skies Through Innovation ... 21
    Inset: Flying By the Rules ... 26
    The Next Generation - Models 24 and 25 ... 26
    High-Flying History Makers ... 28
    Conclusion ... 30
    Key Takeaways ... 32

Chapter 4 Flying to New Heights ... 33
    Introduction ... 33
    Building a Higher-Flying Learjet - Models 28 and 29 ... 34
    Reaching New Heights with the 35 and 36 ... 35
    The Spacious Model 55 and Upgrades ... 36
    Model 31 — The Runt of the Litter Turned Champion ... 38
    The Sky's the Limit with the 31A ... 39
    Pushing Performance with the Model 60 ... 41
    Computer-Aided Design Revolutionizes the Model 45 ... 44
    Legacy of Continuous Innovation ... 46
    Inset: How Stick Pushers Prevent Deep Stalls ... 48
    Key Takeaways ... 49

Chapter 5 Learjet Special Missions — 51

    Introduction — 51
    Movie Magic in the Sky — 52
    Chasing the Sound Barrier — 55
    Inset: A Deadly Dance Gone Wrong — 56
    NASA's Lightning-Fast Lab — 57
    Adversary in the Sky — 62
    Eyes in the Sky — 67
    Unleashing Learjet's Wild Side — 69
    Conclusion — 73
    Key Takeaways — 75

Chapter 6 Learjets Take Hollywood — 77

    Introduction — 77
    Lacy's Famous Clientele — 78
    Fueling the Stars: Learjets Take Off in Hollywood — 79
    Howard Hughes' Last Flight — 82
    Get Ready for a Detour — 84
    Key Takeaways — 85

Chapter 7 Learjets Off the Beaten Path — 87

    Introduction — 87
    Learjets' Shady Side — 87
    Odd Jobs: Learjets' Unconventional Missions — 92
    Conclusion — 94
    Key Takeaways — 96

Chapter 8 Learjet: The Lost Tales — 97

    Introduction — 97
    Mike Abla's Shakey Parlor — 98
    The Mystery of the Barrel Roll Buzz — 100
    Just Reverse the Hook — 101
    The Test Flight on the Edge of Disaster — 102
    A Cat Steers the Pilot — 103
    Flying Cowboys Meet Their Doom — 105
    The Miracle on Lake Michigan — 106
    The Chicken Chucking Blow-Up — 109
    Experience is the Best Teacher — 110
    Key Takeaways — 112

Chapter 9 Final Frontiers — 113

    Innovative Upgrades: 70 and 75 Models Soar to New Heights — 113
    Lear's Winged Wonders - The Learstar and Lear Fan — 115
    Legends Falter - The Ill-Fated 85 — 116

    Learjet's Final Countdown 118
    Key Takeaways 120

Chapter 10 Dreaming of Flight **121**

    Pre-Flight 121
    Cleared for Takeoff 123
    Bill Lear's Big Dream 124
    Against the Wind 126
    Touching the Sky 126
    Imagine That 127

Personal Reflections **129**

References **132**

Chapter 1
Introducing the Learjet Adventure

Welcome!

Hey there, future sky explorers! Have you ever looked up at the sky and wondered how people managed to soar among the clouds like birds?

Well, get ready to embark on an exciting journey through *The Age of Learjet: Adventures, Innovations, and Sky-High Dreams.* In this amazing book, you'll discover how brilliant minds like Bill Lear and Dr. Hans-Luzius Studer turned their wildest dreams into reality, changing the way we see the world from above!

Imagine a time when the sky wasn't just a vast blue ceiling above us but a boundless frontier waiting to be conquered. Back in the day, flying was a daring adventure, reserved for only the bravest. But then came a bunch of visionaries who had a different idea. They wanted to bring the sky closer to us, to make it an open playground for everyone!

Figure 1: A collage of photographs showing different Learjet uses, including military, research projects, and aerobatics.

Picture sleek jets like those in Figure 1 soaring through the air, breaking the sound barrier and whisking people to faraway places in no time. These jets weren't just any planes – they were Learjets, and they were the result of incredible minds working together to push the limits of what was possible. But it wasn't just about the jets; it was about the dreams that took flight alongside them.

In *The Age of Learjet*, you'll uncover stories of determination, innovation, and the firm belief that the sky had no limits. You'll peek behind the scenes and learn how these amazing aircraft were brought to life, from pencil sketches to the very first test flights.

So, get ready to soar through the pages of this book, as we explore the minds of those who dared to dream big, the breakthroughs that changed the course of history, and the incredible journeys that took flight.

From the early days of flight to the jet-setting adventures of today, *The Age of Learjet* will show you that with a little imagination and a lot of determination, even the sky isn't the limit – it's just the beginning!

Buckle up, young explorers. Our journey is about to take off, and the sky is not the limit anymore – it's our playground of dreams!

Inset: Discover Innovation

Idea: Innovation is like being an explorer of ideas. It's about thinking outside the box, dreaming big, and coming up with new ways to remove friction.

One of the best examples of innovation is the wheel — they sure make taking off on land a lot easier.

But friction isn't just physical friction. It can also come in the form of unhappy emotions like fear, distrust, and boredom. Or it could be the time and effort to do something.

So, innovation is finding new ways to make things faster, safer, more reliable, easier, or more fun.

Action: Be an innovation explorer! As you read, pick out ways Learjet removed friction for its customers.

Then, come up with ways you can make some things you do faster, safer, more reliable, easier, or more fun.

Remember, the sky's the limit!

Key Takeaways

- ✓ In the early days of aviation, flying was dangerous and only for daredevils.

- ✓ A group of smart people had a vision to make flying easier and more accessible.

- ✓ They created the Learjet - a sleek, fast jet that could take people long distances quickly.

- ✓ The Learjet made business travel faster and luxury personal travel possible.

- ✓ The creation of the Learjet opened up the skies to more people.

- ✓ With imagination and determination, innovators like the Learjet team made rapid air travel a reality.

Chapter 2
Sky-High Dreams

Introduction

The story of Learjet begins with big dreams and bold imagination. We introduce Swiss founders Studer and Lear who envisioned a swift new jet to race above the clouds.

The Need for Speed

At the time, Lear had converted some old propeller planes into fancy corporate planes called Learstars.

Figure 2: Helmut Horten's Lockheed Model 18 Learstar (Hamel and Park, 2022, 18)

But they were slow, topping out at just 300 miles per hour. Lear knew that the new jet engines being developed could go way faster. He just needed the right plane design to make his dreams take off.

Luckily, an ingenious engineer halfway across the world held the key to realizing his jet-powered dreams.

Meet the Mavericks

Have you ever dreamed of flying your own jet? Back in the 1950s, two men shared that dream: Bill Lear and Dr. Hans-Luzius Studer (prounounced "Hahns-Loo-zee-us Stooder.")

Figure 3: William "Bill" Powell Lear, Sr. (Credit Learjet) (Hamel and Park, 2022, 19)

Lear was a creative inventor living in Switzerland. He imagined a world where businessmen like him could zoom across the country in sleek, swift private aircraft.

Dr. Studer was a talented aeronautical (airplane) engineer, who also lived in Switzerland. He had designed fighter jets for the Swiss Air Force.

Together these bold dreamers would team up to invent the future of business jet travel. Lear had the big dreams and Studer had the engineering talents. But it wouldn't be an easy flight - bumpy skies lay ahead!

Figure 4: Left to right — Dr. Studer with FFA test pilot Bardill, test pilot Bill Lear, Jr., FFA Chief Engineer Spalinger and aerodynamicist Weichelt (Hamel and Park, 2022, 17)

A Brilliant New Partner

Dr. Hans-Luzius Studer was an ingenious aeronautical engineer also living in Switzerland. He had designed advanced fighter jets, like the P-16, for the Swiss Air Force. But the government decided not to buy the P-16 before it went into full production.

Figure 5: Flug- & Fahrzeugwerke Altenrhein P-16 (Hamel and Park, 2022, 9)

Eager to use his talents after this unfortunate setback, Studer soon met inventor Bill Lear. Lear shared his vision for creating a speedy business jet. Studer suggested basing the business jet design on his proven P-16 fighter jet wings. This would allow for a lightweight yet high-performance aircraft - perfect for executive travel.

United by a dream, Lear and Studer quickly became excited partners, combining technical and creative strengths. Their collaboration would pave the way for the future of aviation.

Together they would transform aviation, but the flight path ahead would be anything but smooth.

Lear and Studer made an unstoppable team on paper, but challenges loomed that would test their partnership in the skies.

Dr. Studer's Ingenious Pivot

After the P-16 fighter jet project was canceled, the brilliant engineer, Dr. Studer was determined to channel his fighter jet brilliance into jet breakthroughs, no matter the obstacles.

Along came Bill Lear with his dreams of a fast business aircraft.

At first, Lear planned a turboprop airplane with rear propellers and boundary layer control based on August Raspet's research.

Raspet had told Lear that using boundary layer suction, a modified XV-11A MARVEL aircraft could cruise at 344 mph at 25,000 ft with 1,520 mile range.

Figure 6: XV-11A MARVEL (Hamel and Park, 2022, 19)

It was a bold plan. Aviation was risky in those days. Brave pilots often lost their lives pushing progress forward.

But Dr. Studer saw dangers in the design. During meetings at Lear's Swiss home, Studer tried to convince Lear that the concept was too risky.

Sadly, Dr. Studer's concerns about the design proved true. In 1960, Raspet crashed and died testing a boundary layer control plane.

With his first idea crushed, Lear turned to Studer to get back on track.

Lear planned to design the jet from scratch. But Dr. Studer had a better plan. He convinced Lear to use elements of his P-16 fighter as the basis for this new jet. This saved time and money by building on proven technology.

The P-16 had strong, stubby wings perfect for a light business jet. Dr. Studer took the wings and got busy creating Lear's vision. His disappointment turned into excitement!

Let Your Mind Soar

Now Dr. Studer had a new challenge - turning his bold fighter jet into an airplane for business travel! Soon the drawings took shape into the SAAC-23 design. SAAC stood for Swiss American Aviation Corporation, the company Lear started. The SAAC-23 design was small but mighty - built for speed.

Figure 7: SAAC-23 Low-Speed Wind Tunnel Model (Hamel and Park, 2022, 2)

The pair pictured rich executives flying the SAAC-23 to meetings across the country. No more long bumpy flights in propeller planes! Their jet would whisk travelers along at over 500 miles per hour. Oh, what adventures executives would have soaring high above the clouds!

Bumpy Skies Ahead

Lear and Studer ran into bumpy skies trying to build the SAAC-23. First, the partners planned to make the plane parts in Switzerland and Germany. But there were too many mix-ups and disagreements. The factories did not speak the same language!

With different languages and work styles, there were too many mix-ups. Parts were delayed. Instructions got confused. Everyday disagreements turned into big problems.

Simple things like using different measurement units caused headaches. The Europeans used centimeters while Lear's US team used inches.

All the little cultural differences added up fast. The ocean between the countries made it hard to smooth things out. Lear and Studer realized their global plan wasn't going to fly.

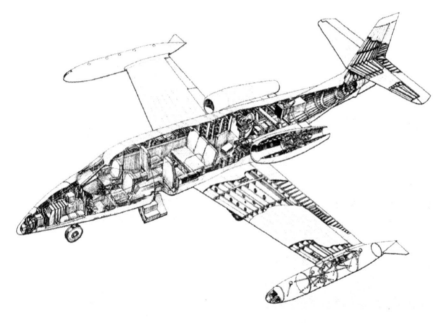

Figure 8: SAAC-23 sectional drawing (Hamel and Park, 2022, 21)

With all these mix-ups piling up, Lear was losing money fast on his dream project. Over 14 million dollars invested, and still no finished jet to show for it!

Dealing with confusing foreign regulations was the last straw. There was only one way to get his jet dreams safely back on track.

Kansas or Bust

Frustrated, Lear decided to move production across the ocean to Wichita, Kansas in 1962. This meant redoing all the tooling and training new builders. It was a big risk, but Lear kept chasing his jet dreams.

Studer stayed behind in Switzerland. But his clever wing design traveled with Lear. Through it all, Lear kept picturing executives flying high over the USA in the SAAC-23. That dream kept him going, even when skies grew cloudy.

Wheels Up!

Finally, in 1963 the first SAAC-23 prototype was ready. Lear renamed it the Learjet Model 23. Just like he dreamed, it was small and speedy. Seeing the first Learjet filled Lear with pride. His dreams were now real!

Figure 9: Lear Jet 23 first flight on October 7, 1963 (Credit Learjet) (Hamel and Park, 2022, 43)

On October 7, 1963, the very first Learjet plane took to the open skies over Kansas. By February, 1964, test pilots pushed the Learjet to Mach 0.905 (609 mph). It was the fastest business jet in the world.

On July 31, 1964, the Learjet earned official approval from the Federal Aviation Administration. Now Lear could begin full production of his game-changing jet. His SAAC-23 dream plane was officially the Learjet 23 reality!

The first flight of the Model 23 showed Lear never gave up on his sky-high dream.

Dreams Take Flight

Lear never gave up on his sky-high dream of building a fast business jet. He imagined it, believed in it, and made it real despite setbacks along the way. With inspired partners like Studer, Lear turned his imaginings into inventions that made aviation history.

The first Learjet combined creativity, innovation, and the thrill of soaring among the clouds. It gave business travelers a new world of possibilities. And it all started with the power of a dream.

Lear's bold vision launched a future filled with sleek jets whisking people across the country in comfort and style. The Learjet changed business travel forever.

Its success continues to inspire new generations to dream big – then design and build to make those dreams take flight!

Figure 10: Lear Jet 23-003 — the first business jet in history to be delivered to a customer (*Credit Jaap Niemeijer*) (Hamel and Park, 2022, 48)

Inset: What is "Boundary Layer Control"?

Have you ever gone down a big, fast water slide? As you zoom down, the water rubbing the slide slows down. That slow water is like a "boundary layer" hugging the slide.

It's the same with airplanes. A thin layer of slow air forms along the wing. This boundary layer air creates drag that holds the plane back, like slow water slowing down your slide.

Some smart engineers tried to fix this using suction. They put tiny holes in the wings to suck away the slow boundary layer! It's like if your slide had holes to drain the slow water, so you could go faster.

The holes make the plane more efficient, like clearing leaves from your slide. But just like a blocked slide, blocked holes can cause problems in flight.

You'll hear more about the boundary layer later in this book.

Key Takeaways

- ✓ Bill Lear and Dr. Hans-Luzius Studer both dreamed of creating a revolutionary swift new aircraft despite facing numerous obstacles.

- ✓ Studer convinced Lear to base their jet design on proven technology from his fighter jet wings, pivoting after a setback.

- ✓ The partners combined Lear's imaginative vision with Studer's technical expertise to sketch the first SAAC-23 jet plans.

- ✓ Manufacturing the SAAC-23 overseas proved problematic due to frustrating language and cultural barriers.

- ✓ Lear took a risk moving production to Wichita, Kansas in 1962, determined to make his jet dreams reality.

- ✓ The SAAC-23 rolled out in 1963 as the pioneering Learjet Model 23 and soon reached record speeds during test flights.

- ✓ Imagination, innovation, and perseverance allowed Lear and Studer to overcome challenges and achieve aviation history.

Chapter 3
Pushing the Limits

Introduction

Soar into aviation history as we look back on the trailblazing early Learjets! When these speedy jets first took to the skies in the 1960s, no one expected such small aircraft could shatter speed records across America.

Pilots like Clay Lacy pushed the Learjet 23 to over 600 mph, shocking crowds with the jet's breathtaking velocity. But danger often lurks at the edge of innovation. Excitement turned to tragedy when early crashes tarnished the young jet's image.

Now Learjet faced its biggest test - proving their aircraft was safe enough to earn back public trust. Through relentless dedication, Learjet engineers responded to each setback with critical safety advances, steadily rebuilding confidence in jet travel.

Later models like the 24 and 25 climbed even higher, cruising smoothly above congested airspace. By achieving rigorous FAA certification, Learjet cemented its reputation for performance and reliability.

Come aboard as we relive how these daring early Learjets made aviation history. Despite daunting trials, their spirit of innovation and persistence sparked a revolution in executive travel. The sky was no longer the limit!

Blazing a Trail with the Speedy 23

The early Learjet 23 didn't just carry business travelers - it rocketed them across the skies at record-breaking speed. Soon after its first flight in 1963, the Jet Age dream machine went on to blaze an aviation speed trail.

Figure 11: Lear Jet 24-100 returns from record setting global flight. From left aviation journalist J. Zimmermann, Lear Jet VP B. Sipprell, Pilot John Lear, President Bill Lear, PIC H. Beaird, Moja Lear and Pilot R. King (Credit Learjet) (Hamel and Park, 2022, 52)

In May 1965, pilots Clay Lacy and Jack Conroy pushed the Learjet 23 to fly from Los Angeles to New York and back again in just 10 hours and 21 minutes. That sliced a thrilling 5 hours off the previous coast-to-coast record.

Figure 12: Record Lear Jet 23-012, Clay Lacy and General James Doolittle 1965 (Doolittle is best known for leading the 'Doolittle Raid' as a Lieutenant Colonel, a daring B-25 raid on Tokyo during World War II.) (Hamel and Park, 2022, 18) (Hamel and Park, 2022, 50)

But danger often lurks when daring new frontiers are crossed. While the Learjet 23 thrilled crowds with its blazing speed records, behind the scenes, troubling signs appeared.

Minor incidents whispered of risks that could dash the dreams of business jet travel. Learjet engineers were about to face their biggest challenge yet – earning passengers' trust that flying in a tiny jet was safe. Their persistence would be tested like never before after tragedy struck the pioneering Learjet.

Other daring pilots like Hank Beaird further pushed the limits. By February 1964, Beaird flew the Learjet 23 to a breathtaking Mach 0.905 - over 600 mph. No business jet had ever topped those speeds.

Figure 13: In-flight photo of the instrumented Lear Jet 23-001 prototype (Hamel and Park, 2022, 44)

These heart-pounding early speed runs opened the world's eyes to the Learjet's immense potential. They sparked imaginations about what business travel could be. Roaring over the landscape below, the Learjet set speed records that still stand today. Those flights trailblazed a future where executives could jet across the country in half a day.

The Learjet 23's high-flying achievements stretch far beyond mere records. They represent the adventurous spirit of pushing boundaries that defined early Learjet innovations.

But how long could the Learjet's luck hold? Even trailblazers face trials by fire. Just when it seemed nothing could stop the swift Learjet's meteoric rise, disaster struck. Cracks formed in the jet's breakneck success. Now Learjet's real test began - could they pick up the pieces and rebuild public trust? Or had they pushed their limits too far?

The answers would rely on Learjet's unrelenting dedication to safety innovations. Their most important work was just getting off the ground.

Tragedy Sparks Safer Skies Through Innovation

The excitement of speed records was dampened by sobering early accidents, including the first Learjet 23 prototype crashing in 1964. Learjet engineers responded with critical safety innovations so passengers could trust in jet travel.

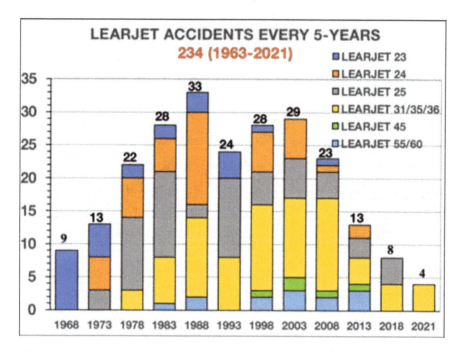

Figure 14: Learjet accidents every 5 years (Hamel and Park, 2022, 54)

Tragically, more accidents in 1965 showed Learjet's work was not done. In October 1965, a Learjet Model 23 crashed, killing both pilots. Investigators blamed an electrical failure, but the exact cause remained unclear.

In response, Learjet added backup attitude indicators. The modifications ensured that backup the system stayed powered.

But more accidents followed, including another with two fatalities in 1966. Disturbingly, Learjet's fleet accidents kept rising, averaging over 4 per year by the late 1980s.

Learjet refused to accept the grim trend. After each tragedy, engineers dove into understanding the cause. They steadily enhanced safety systems, determined to protect future pilots.

New backups for attitude indicators prevented electrical failures from turning deadly. Redundant sensors helped pilots stay oriented if primaries failed.

Most importantly, Learjet kept learning from each setback. Their culture of tirelessly improving safety paid off. By the 2000s, accident rates finally curved downward. New Learjet designs became progressively safer.

Figure 15: Lear Jet 23/24 Assembly line in Wichita (1966) (Hamel and Park, 2022, 45)

One major improvement was switching to meet FAA Part 25 airworthiness standards in 1965. Part 25 standards are far more stringent than the previous certification. This demanding process required upgrades like redundant systems and a "stick shaker" to warn pilots of imminent stalls.

Lear's relentless passion for safety was legendary. When testing a new stronger Learjet windshield, Bill Lear personally jumped up on the glass as engineers watched! While unconventional, Lear's hands-on methods reflected the company's intense commitment to safety.

Figure 16: Lear Jet windshield cracker, Bill Lear (Hamel and Park, 2022, 55)

On another occasion, Lear planned to sit behind a windshield during a bird strike test. But just before the test, he was called away for an urgent phone call. While Lear took the call, the a 4-pound chicken blasted through the windshield. See "The Chicken Chucking Blow-Up" in Chapter 8.

But Lear wasn't deterred - he used the setback to improve the design. His personal approach highlighted his dedication to safety innovations.

Figure 17: Learjet bird strike gun test stand (Credit Dick Kovich) (Hamel and Park, 2022, 327) © 2023 The Authors, under exclusive license to Springer Nature Switzerland AG, reproduced with permission

Rather than hiding setbacks, Learjet highlighted them to guide new safety advances. Their steady modifications after each incident pushed engineering limits to help prevent future tragedies.

Learjet's culture of tirelessly enhancing safety laid the foundation for generations of jets. Their continual progress instilled confidence in executives hesitant to fly in early business jets. Unwavering persistence and innovation transformed air travel into the trusted mode it remains today.

Thanks to Learjet's critical safety improvements, accidents started to decline. This restored confidence, allowing Learjet to focus on developing more advanced models.

Inset: Flying by the Rules

Pilots don't just need skills - they need to know aviation rules and regulations.

In the early years, Learjets were certified under Civil Aviation Regulations (CARs). These were like rulebooks from the Civil Aeronautics Authority, aviation's early overseeing body.

By the 1960s, CARs were replaced by Federal Aviation Regulations (FARs). These new strict guidelines were issued by the newly formed Federal Aviation Administration (FAA).

FAR Part 25 set stringent airliner-level standards for aircraft design and performance. When the Learjet Model 24 met Part 25, it proved worthy of professional crews.

Thanks to FARs, the Learjet models that followed had to prove their mettle to fly executives with confidence. Just like pilots, those early Learjets trained hard to earn their wings!

The Next Generation - Models 24 and 25

Building on safety advancements, Learjet engineered its next generation of jets to climb higher and faster. The Model 24 first flew in 1966 with upgraded engines and stronger pressurization to reach new altitudes.

Figure 18: Lear Jet 24 (24-123) (Credit FlightAware) (Hamel and Park, 2022, 57)

But the Model 24 will forever be known for another milestone – becoming the first business jet to achieve FAA Part 25 certification. Meeting these stringent airliner standards demonstrated the Learjet was robust enough for professional crews and serious travel.

Figure 19: Lear Jet 25 (Can you spot the differences between a Model 24 and Model 25?) (Hamel and Park, 2022, 57)

The slightly stretched Model 25 followed a year later in 1967. With 51,000 feet now approved, the Models 24 and 25 could cruise comfortably above weather and congested airspace.

Certifying the Learjet to Part 25 was a crowning achievement in proving it was here to stay. No longer just a daring upstart, Learjet cemented its reputation as a mature jet built to the highest standards.

Executives could stay productive crossing the country above traffic and turbulence. The smooth rides let corporations conduct business at jet speed. Learjet's next generation carried on the relentless quest to open new possibilities in the sky.

High-Flying History Makers

Gates Learjet made headlines in the 1970s and beyond with new milestones. In 1976, a patriotic Learjet 36 set an around-the-world record, flying 23,000 miles in just over 57 hours.

Figure 20: World record-breaking Learjet 36-014 (N200Y) (Credit Learjet) (Hamel and Park, 2022, 77)

Topping even airliners, Learjets were approved to soar at 51,000 feet starting in 1977. The Boeing 747, for example, normally cruised between 31,000 and 38000 feet.

Famed astronaut Neil Armstrong pushed a Learjet 28 to new heights in 1979, setting three world records.

Figure 21: Chief Engineers Hank Waring and Don Grommesh (Credit Wichita State University & Grommesh) (Hamel and Park, 2022, 46)

"Aviatrix" Brooke Knapp flew into the record books by breaking 100 records in Learjets, including a 50-hour round-the-world trip. Her feats were bested in 1996 by a souped-up Learjet 35A.

From blazing speed to extreme altitude, Gates Learjet pilots pushed boundaries. Their daring adventures carved out new possibilities for business aviation.

Conclusion

The early Learjet models stand today as aviation pioneers that pushed boundaries in business travel. Where limits existed, Learjet innovations blew past them.

The Model 23 blazed transcontinental speed records that made coast-to-coast jet travel a thrilling reality. Later models like the 24 and 25 climbed to heights where airliners feared to fly.

When tragedy struck, Learjet persisted in creating a new standard for safety. Their pursuit of FAA Part 25 certification cemented the Learjet as a trusted travel investment.

Most of all, the company proved that small didn't mean limited. Their early jets consistently exceeded expectations of what business aircraft could accomplish.

Figure 22: On September 10, 2015, a Learjet 75, flanked by the original Learjet 23, sets a world speed record across the U.S., marking the 50th

anniversary of Clay Lacy's 1965 record-setting transcontinental round-trip flight. Pilot Jeff Triphahn, Clay Lacy and Pilot Ed Hillis (Credit Learjet) (Hamel and Park, 2022, 51)

Speed, safety, performance – the early Learjets moved aviation forward on all fronts through sheer determination. In the process, they opened executives to a world of time-saving possibilities.

By proving its mettle, the pioneering Learjet earned a place in history. Its success cleared the skies for the flexible business travel millions enjoy today. Learjet invented the future of aviation – then dared rivals to keep up.

Key Takeaways

- ✓ The early Learjet 23 broke speed records, rocketing across the country faster than ever before. This opened up new possibilities for business travel.

- ✓ Tragedy struck with early crashes, but Learjet responded with critical safety innovations to rebuild trust. Their persistence set a new standard.

- ✓ Upgraded Learjet Models 24 and 25 were certified to stringent FAA Part 25 airliner standards, proving their robust design.

- ✓ Learjets kept achieving aviation firsts, from reaching new altitudes to record-breaking global flights. Pilots pushed boundaries.

- ✓ Learjet's culture was defined by tirelessly enhancing safety and exceeding expectations. They overcame setbacks through determination.

- ✓ By pioneering speed, safety, and performance, Learjet cemented its reputation for trusted travel. The company invented the future of business aviation.

- ✓ Each innovation opened new possibilities in the sky. Learjet blazed trails that shaped modern executive travel.

Chapter 4
Flying to New Heights

Introduction

Let's jet forward to the daring Learjets that took executive travel to new heights!

While past models blazed speed records, these next-generation jets soared to the stratosphere and beyond. The advanced Models 28 and 29 climbed effortlessly above congested airspace, certified to cruise at 51,000 feet.

Later, the groundbreaking Model 60 pushed higher and faster with its powerful engines and specially designed wings. This cutting-edge jet sliced through the skies with ultra-efficient aerodynamics mapped by supercomputers.

The revolutionary Model 45 became the world's first business jet engineered completely on computers. By pioneering digital design, Learjet propelled aviation into the future.

But reaching new heights didn't come easily. It took visionary thinking and tireless refinement to push the limits. Each leap forward built on lessons learned from past setbacks and tragedies.

Join us on an inspiring journey through Learjet's relentless pursuit of excellence. Their daring innovations raised the bar for an entire industry. With each new model, Learjet uplifted possibilities in private air travel. The sky was no longer the limit!

Building a Higher-Flying Learjet - Models 28 and 29

While racing across the landscape, Learjet had its sights set on reaching new heights. The company unveiled the advanced Models 28 and 29 in 1978, certified to soar up to 51,000 feet.

Figure 23: Historic Learjet 28-001 wing — the first production jet with winglets (Credit R. Randall Padfield) (Hamel and Park, 2022, 71)

Built for smooth high-altitude cruising, these new jets featured more powerful engines and increased fuel capacity. Their optimized aerodynamics reduced drag at the high speeds required in the stratosphere.

In February 1979, famed astronaut Neil Armstrong put the Model 28's ceiling to the test. Along with Learjet pilot Pete Reynolds, Armstrong flew a Model 28 to an astonishing 51,130 feet, setting three world records.

Figure 24:: Armstrong and Reynolds during historical flight. Learjet 28-001 en route to Armstrong Museum (Credit Learjet) (Hamel and Park, 2022, 79)

With Models 28 and 29, Learjet seized dominance of the upper atmosphere for business travel. Executives could fly coast-to-coast in the stratosphere, avoiding bad weather and crowded air corridors below. Once again, Learjet opened new possibilities in the sky.

Reaching New Heights with the 35 and 36

In the 1970s, Learjet soared to new heights with the cutting-edge Models 35 and 36. These jets weren't just advanced – they were cosmic!

Figure 25: Raisbeck ZR LITE demonstrator Learjet 35A (Hamel and Park, 2022, 89)

While past Learjets cruised at 41,000 feet, the suped-up Models 35 and 36 rocketed to 51,000 feet with turbofan power. They flew so high they could almost glimpse outer space!

To pilot the thrilling new Models 35 and 36 was to skim the stratosphere, floating smoothly above the clouds, turbulence, and traffic below.

Learjet stretched these jets longer to carry more fuel for epic nonstop flights. Their pointy noses helped slip through the air at record speeds.

VIP and military planes used the Models 35 and 36 to Jet in style. Even world-famous astronaut Neil Armstrong gave the 35 his stamp of approval.

The Spacious Model 55 and Upgrades

The Learjet Model 55, introduced in 1979, provided more room for passengers compared to previous Learjets.

Figure 26:: Learjet 55 receives FAA Type Certificate on March 18, 1981 (Credit Business Aviation) (Hamel and Park, 2022, 81)

A longer and wider fuselage (the main body of the airplane) allowed comfortable seating for up to eight flyers. More powerful TFE731 engines gave the Model 55 impressive performance.

The Model 55 also utilized the winglets first introduced on the Models 28 and 29. These wingtips improved fuel efficiency and climb rates.

Figure 27: Delta fins are the small triangular fins on either side of the tail. They help stabilize the Learjet in flight by smoothing airflow. The fins' triangular shape also reduces drag, helping models like the 36A (pictured) and 55C cut through the sky more efficiently. The delta fins may look small, but they make a big difference in flight! (Hamel and Park, 2022, 70)

In 1986, the Model 55B brought extended range capabilities and a glass cockpit with advanced avionics.

Then in 1987, the Model 55C appeared with further engine and aerodynamic refinements, including delta fins.

Together the Model 55, 55B, and 55C delivered space, comfort and high-altitude cruise speeds in a proven airframe.

Upgrades like winglets and digital displays moved business aviation forward, while retaining Learjet's hallmark performance. The Model 55 series stretched possibilities for executive travel.

Model 31 — The Runt of the Litter Turned Champion

People who bought Learjets really liked how the Model 35 performed. But they wished it was easier to control during flight and could land more slowly. So, in the iconic 1980s, Learjet's tubular Model 31 combined the best – a Model 35 body with groundbreaking wings.

Figure 28: Learjet Model 31 (Credit FlightAware) (Hamel and Park, 2022, 81)

The Model 31 used the proven Model 35/36 fuselage paired with new wings and winglets from the Model 29. This wing swap improved handling thanks to the addition of delta fins on the tail.

But the Model 31's short range and outdated avionics made it a dud with buyers. Learjet had struck out trying to upgrade this blast from the past.

Or had they? Just when it seemed the Model 31 was headed for the moon in a bad way, Learjet pulled off a cosmic comeback.

The Sky's the Limit with the 31A

While the original Learjet Model 31 had limited range, the company saw an opportunity to transform this laggard into a leader. In 1990, they unveiled the revamped Model 31A, which took the best of the old and made it new again

Figure 29: Learjet 31A (Hamel and Park, 2022, 116)

Now certified to soar at 51,000 feet, the Model 31A climbed above congestion and weather. Digital avionics and redesigned controls also boosted performance.

Later enhancements allowed even higher speeds, takeoff weights, and range. An Extended Range 31A variant could fly even farther without refueling.

With its blend of proven reliability and next-gen upgrades, the Model 31A became a customer favorite. It delivered an unbeatable balance of comfort, capability, and operating costs.

The Model 31A showed how Learjet turned setbacks into comebacks. By listening to feedback and optimizing designs, they took a model that missed the mark and transformed it into a bullseye.

Pushing Performance with the Model 60

Learjet continued its ambitious ascent throughout the 1990s. The company unveiled the groundbreaking Model 60 - an ultra-efficient jet that climbed faster and flew higher thanks to its powerful new engines.

Figure 30: Learjet Model 60 received FAA type certificate on January 15, 1993 (Hamel and Park, 2022, 118)

The Model 60 also had a cabin stretched nearly two feet longer, giving passengers more legroom. Its smoothed aerodynamics, from the redesigned wings to sleek engine mounts, allowed greater fuel economy.

Learjet dreamed up a revolutionary new wing design for the Model 60 using futuristic technology – supercomputers from NASA!

NASA's powerful machines helped shape the most precisely contoured wings ever. It was like having a rocket scientist do math homework. The computers crunched endless calculations to optimize every curve.

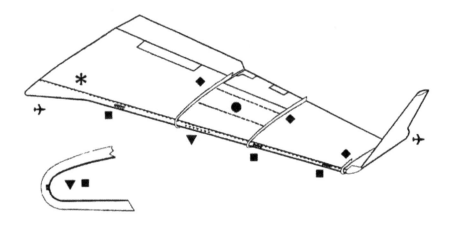

Figure 31: Learjet Model 60 wing devices (Hamel and Park, 2022, 118)

Testing in wind tunnels fine-tuned the space-age design. Tiny vortex generators near the wingtips created swirling air over the wing to stir up the boundary layer – yes, that pesky boundary layer from Chapter 2!

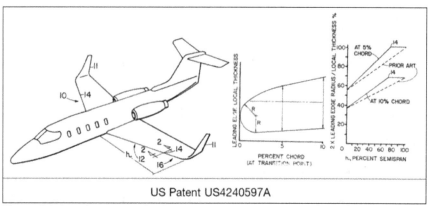

Figure 32: Learjet patent for boundary Layer Energizers (BLEs) mounted on the wing (Hamel and Park, 2022, 76)

It was wings like no Learjet before! The cutting-edge shape let the Model 60 slice through the skies faster while burning less fuel. Engineers beamed with pride at their sleek new creation.

The sky-high-tech wings were just one amazing advance that made the Model 60 a next-gen Learjet. This jet of the future flew passengers higher in first-class comfort.

For pilots, the advanced Model 60 provided state-of-the-art avionics and handling. Breakthroughs like a computer-assisted rudder boost system reduced workload and made flying easier.

Figure 33: Learjet 60 cockpit design (Credit Carlos Alberto Rubio Herrera) (Hamel and Park, 2022, 117)

The pioneering Model 60 entered service in 1993 as the most thoroughly modernized and capable Learjet ever conceived. With unparalleled performance and luxury, Learjet once again soared ahead of the pack.

Computer-Aided Design Revolutionizes the Model 45

The Learjet Model 45 was a game-changer in business jet design. While past Learjet models pushed performance limits, the Model 45 soared to new heights as the first Learjet engineered completely on computers.

The Model 45 was also a global first - the first Learjet designed and built across multiple countries. Bombardier worked with de Havilland Aircraft in Canada and Short Brothers in Northern Ireland.

Short Brothers built the fuselage (the main body) in Belfast. De Havilland built the wings in Canada. Then they shipped the parts to the Learjet factory in Wichita, Kansas. There, workers assembled the final airplane.

Figure 34: Workers loading first Learjet Model 45 fuselage for shipment from Belfast, Northern Ireland to Wichita, Kansas in October 1994 (Hamel and Park, 2022, 122)

But coordinating across continents proved tricky. Each company used different and incompatible computer systems!

Unveiled in the mid-1990s, the Model 45 special wings made better using cool NASA computer programs. Experts in aerodynamics (the study of how air moves around things) crafted the wings on high-tech computers instead of paper.

Powerful simulations tested virtual wings in a variety of extreme conditions with ease. The best-performing designs were chosen for real-world testing.

Figure 35: 1/6 scaled Learjet 45 in NASA Langley TDT (Credit Abe Jibril) (Hamel and Park, 2022, 124)

Wind tunnel and flight tests further refined the computer-calculated contours into wings capable of smooth,

Mach 0.81 cruise. Improved controls and spacious cabins completed the high-tech wonder.

Figure 36: Learjet 45XR (Credit Barry Shipley) (Hamel and Park, 2022, 128)

The Model 45 entered service in 1995 as the most thoroughly computerized business jet ever. By pioneering digital design, Learjet propelled aviation firmly into the future.

Legacy of Continuous Innovation

Throughout its evolution, Learjet cemented a legacy of continuously redefining the limits of private flight. Each new model further pushed the boundaries of speed, altitude, comfort, and safety.

But reaching new heights didn't happen easily. It took dedication to precision design, extensive testing, and tireless refinement. Learjet engineers embraced the challenge, driven to make business travel smoother, swifter, and more productive.

At times, progress came in the wake of tragedy, as with early accidents that spurred vital safety advances. But Learjet turned setbacks into motivation to improve.

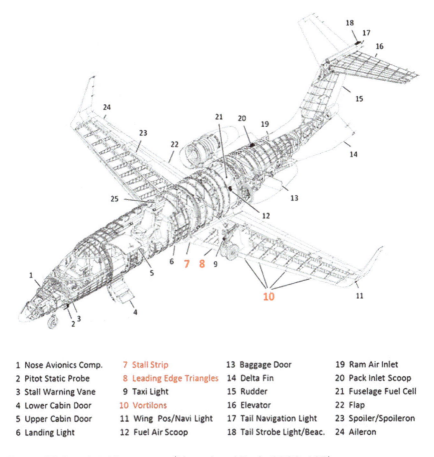

1 Nose Avionics Comp.	7 Stall Strip	13 Baggage Door	19 Ram Air Inlet
2 Pitot Static Probe	8 Leading Edge Triangles	14 Delta Fin	20 Pack Inlet Scoop
3 Stall Warning Vane	9 Taxi Light	15 Rudder	21 Fuselage Fuel Cell
4 Lower Cabin Door	10 Vortilons	16 Elevator	22 Flap
5 Upper Cabin Door	11 Wing Pos/Navi Light	17 Tail Navigation Light	23 Spoiler/Spoileron
6 Landing Light	12 Fuel Air Scoop	18 Tail Strobe Light/Beac.	24 Aileron

Figure 37: Learjet 45 cutaway (Hamel and Park, 2022, 127)

The result was a fleet of jets unmatched in performance, luxury, and reliability. Learjet's commitment to relentless innovation opened the skies to flexible travel for executives worldwide.

By never being complacent, Learjet uplifted an entire industry. Their quest for excellence raised standards for business aircraft generation after generation. Even today, Learjet's spirit of boundary-pushing progress remains alive.

Inset: How Stick Pushers Prevent Deep Stalls

A stall happens when a plane's wings get too steep of an angle of attack (the angle of the wings through the air). This disrupts the airflow and makes the plane suddenly dive downward.

Normally pilots fix stalls by adjusting the controls on the tail to lower the angle of attack.

But on some planes like Learjets, the tail can get stuck in the turbulence (messy airflow) from the stalled wings. The turbulence air makes it really hard for pilots to recover, leading to a dangerous "deep stall" and risk of a crash.

To prevent deep stalls, some planes have a stick pusher. This automatically pushes the control stick forward when the angle gets too steep.

The stick pusher helps avoid a deep stall lowering the nose before it's too late.

Key Takeaways

- ✓ The advanced Learjet Models 28 and 29 could fly higher than ever, certified up to 51,000 feet in the stratosphere above weather and traffic congestion.

- ✓ Famed astronaut Neil Armstrong set three world records in a Model 28, reaching an astounding 51,130 feet altitude near the edge of space.

- ✓ The 1970s Models 35 and 36 rocketed to new heights of 51,000 feet with more powerful turbofan engines, allowing smoother high-altitude cruising.

- ✓ The 1990s Model 60 featured stretched cabins and computer-optimized wings for greater comfort, speed and fuel efficiency at high altitudes.

- ✓ The Model 45 was the first business jet engineered entirely on computers, enabling aerodynamicists to virtually design and test countless wing shapes.

- ✓ Each new generation pushed boundaries in speed, altitude, range, efficiency and luxury through tireless innovation and testing.

- ✓ Learjet turned setbacks into motivation to improve, advancing safety and operability after early accidents revealed areas for enhancement.

- ✓ Extensive wind tunnel and flight testing refined the computer-calculated designs into wings capable of unmatched cruise performance.

- ✓ By never being complacent, Learjet continuously raised the standard for the entire business jet industry as competitors worked to match their capabilities.

- ✓ Their relentless innovation spirit opened new possibilities in private aviation and accessible global travel for business.

Chapter 5
Learjet Special Missions

Introduction

Prepare for a high-octane adventure as we explore some of the most daring and innovative missions flown by Learjets!

Figure 38: Air-to-Air Photographer Paul Bowen sitting at the Tail Gunner Station of a WWII Vintage B-25 bomber shooting a Learjet 45XR (Credit Paul Bowen) (Hamel and Park, 2022, 199)

In this chapter, you'll learn how Learjets starred in Hollywood blockbusters, raced to break aviation records, conducted scientific research, and much more. From the big screen to the frontiers of flight, these agile jets carried out special operations most civilian aircraft never could.

Learjets starred on the silver screen, carrying innovative cameras that captured thrilling aerial scenes in Hollywood blockbusters. How did these versatile jets help film high-flying action movies?

Movie Magic in the Sky

Figure 39: Clay Lacy demonstrating his Astrovision™ Panorama Periscope Camera (Hamel and Park, 2022, 202)

Lights, camera, action! Learjets played a starring role in revolutionizing Hollywood movie-making and aerial filming. Famous pilot Clay Lacy outfitted his Learjet 25 with an innovative camera system called Astrovision. It had periscopes and cameras that could rotate 360 degrees and film the action happening all around the plane.

Before Astrovision, photographers had to mount individual cameras pointing in different directions to capture scenes during flight. But Lacy's dual periscope system allowed panoramic filming from inside the Learjet.

Now cameras no longer need to hang out of windows!

Figure 40: Learjet 35A Nose Camera (Hamel and Park, 2022, 201)

During the 1970s and 80s, Lacy used his Astrovision-equipped Learjet to film some of Hollywood's most memorable aerial action sequences. Car chases, jet dogfights, and other dramatic scenes were captured from Lacy's camera-ready Learjet. He even shot commercials for airlines and the aircraft industry.

Lacy's Astrovision brought movie magic to the sky. The cameras could swivel rapidly to follow high-speed jets making tight turns and other daring moves. The Learjet itself was fast and agile enough to keep up with the action. Lacy captured exhilarating footage that would have been impossible from other aircraft or on the ground.

Figure 41: Learjet 25 joins French Mirages F1 (Credit Clay Lacy Collection) (Hamel and Park, 2022, 202)

So the next time you watch a breathtaking aerial scene in a blockbuster movie, you may have a versatile Learjet camera platform to thank for those images! Learjets helped revolutionize Hollywood's ability to film action sequences and commercials from the air.

Figure 42: Wolfe Air Learjet 25B showing camera pods mounted under its wings (Hamel and Park, 2022, 203)

Learjets chased down aviation milestones to photograph the fastest jets in flight, though tragedy struck during one famed photoshoot. What happened when a Learjet tried to capture five supersonic jets in formation?

Chasing the Sound Barrier

Learjets were also ideal for capturing photos of aviation milestones and record-breaking airplane formations. Their speed and maneuverability allowed them to chase even the fastest jets of the era.

In 1966, Clay Lacy used Frank Sinatra's Learjet to film a formation flight of five supersonic jets powered by General Electric engines. The plan was to capture an impressive aerial photo shoot. But tragedy struck mid-flight.

One of the jets, an F-104 Starfighter, got caught in the wake turbulence from the giant XB-70 Valkyrie bomber. The F-104 flipped over the Valkyrie and crashed into its vertical stabilizers. Both aircraft went down in flames, with the loss of two test pilots.

The accident cut short what was intended to be a photogenic gathering of the era's most powerful jets. Learjets like the one flown by Lacy had the perfect mix of speed, climbing power, and maneuverability to chase such planes for photos. Their versatility as camera platforms allowed spectacular aerial shots despite hazards and risks.

So while the ill-fated 1966 flight ended in disaster, it showed off the Learjet's talents for tracking milestones in aviation history. No other aircraft had their nimble abilities to chase and photograph the fastest jets of the day.NASA used customized Learjets as flying science labs to gather data and test technologies at the edge of space. What kinds of research missions did NASA's high-flying Learjet fleet carry out?

Inset: A Deadly Dance Gone Wrong

Five supersonic jets cut through the clear blue sky with Frank Sinatra's Learjet 24A. But danger lurked behind the mighty XB-70 Valkyrie bomber.

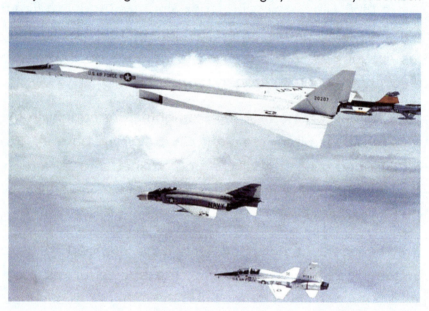

Figure 43: Learjet 24A photo shoot (Credit US Air Force) (Hamel and Park, 2022, 200)

Just as the photoshoot ended, catastrophe struck.

The smallest jet, an F-104 Starfighter, drifted into the powerful vortex of swirling air left behind the giant bomber's wingtips. Buffeted violently, the hapless jet flipped over the back of the bomber, smashing into its vertical tail.

Both damaged aircraft spiraled downward, the F-104 disintegrating into fragments as the Valkyrie struggled in vain to regain control.

The disaster claimed two test pilots' lives that day.

NASA's Lightning-Fast Lab

NASA found Learjets to be ideal high-flying laboratories for scientific research and testing cutting-edge aerospace technologies. The agile jets allowed engineers to conduct experiments and carry specialized equipment to collect data.

Figure 44: Learjet 23-084 (N119BA) noted with NASA colors and a Buzz Aldrin label as seen in 2019 (Hamel and Park, 2022, 151)

Four different Learjet models served NASA over the years. They were modified with sensors and used on missions studying wake vortices (spirals of messy air) left by large planes. This research led to changes reducing dangerous turbulence for aircraft flying in formation.

Figure 45: Lear Jet 23 and USAF T-37 probing B-747 wing tip vortices (Hamel and Park, 2022, 153)

Other NASA Learjets became high-flying ears, listening closely to the roar of jet engines. These flying sound labs recorded the shrieks and rumbles during special noise tests.

Engineers wanted to understand what parts of jet motors made the most racket. How could they design the engines to run quieter and less rumbly?

The Learjets chased after noisy planes with microphones aimed at their loud engines. They gathered data on new engine types to see if they sounded smoother.

Figure 46: Lear Jet 25 with chevron nozzle (ridged end of the jet engine to reduce noise (Hamel and Park, 2022, 157)

Figure 47: Lear Jet 25 chasing NASA's JetStar propeller noise (Hamel and Park, 2022, 157)

Astronomy missions saw Learjets transformed into night-vision space telescopes. Scientists mounted special infrared cameras that could detect heat and faint light.

As the Learjets cruised miles above Earth, the telescopes scanned the darkness. They peered at glowing gas clouds and newborn stars normally invisible to the human eye.

The infrared images from jet black skies revealed secrets of the universe.

Figure 48: Lear Jet 24B observatory and the 12 inch infrared telescope (Hamel and Park, 2022, 155)

NASA's Learjets served as the perfect stable platforms. Their smooth flights enabled the clearest cosmic views.

NASA's researchers also used Learjets to test new technologies before sending them into space. These speedy jets made the perfect flying laboratories.

One important thing researchers tested was new satellite dish antennas. These tiny antennas used special technology that may work in space someday.

NASA attached prototype antennas to a Learjet Model 25. Now the jet could beam signals to and from satellites high in orbit.

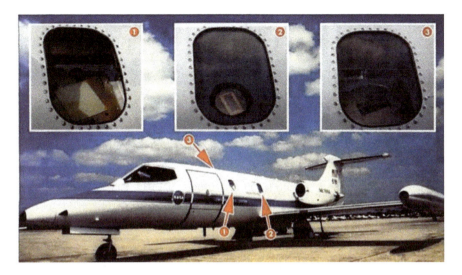

Figure 49: This Learjet Model 25 has three special antennas attached to it. One antenna in the front window was made by Boeing. Another antenna in the back window was made by Texas Instruments. The third antenna on the other side was made by Martin Marietta. NASA (Credit AW&ST) (Hamel and Park, 2022, 159)

On one exciting flight, the jet zoomed through the sky at nearly the speed of sound. An antenna on the rear window called out to a NASA satellite. The satellite sent messages right back to another antenna on the front window.

The test was a big success! NASA saw that the antennas worked great at jet speeds and high altitudes.

Because of these cool airborne trials, NASA started using the tiny antennas in real space missions. The lightweight dishes let satellites communicate better from orbit.

Figure 50: NASA Lear Jet 25 upon retirement (Credit Tom Gabbit) (Hamel and Park, 2022, 159)

NASA used their Learjets many times to test new technologies. The speedy jets helped NASA quickly "try before they fly" when developing new systems for space. Once again, the Learjet showed it could do more than carry business travelers. It helped NASA push technology to new heights!

All this research required aircraft that were reliable, versatile, and capable of high-altitude data gathering. NASA's customized Learjets filled this role as fast and efficient flying laboratories.

When not serving science, Learjets took on another challenging role: adversary.

Learjets serve as "enemy" aircraft, testing the skills of combat pilots and the capabilities of defense systems. How do Learjets help get military air crews battle-ready?

Adversary in the Sky

Learjets take to the skies as "aggressor" aircraft to help train military combat pilots and test battle readiness. Outfitted with special instruments, they provide real-world combat simulations for the latest fighter jets and defense systems.

Figure 51: C-21A Learjet and F-15s during tactical training and mission support (Hamel and Park, 2022, 160)

To challenge pilots' skills, Learjets are equipped as adversaries with electronic jamming pods, missile trackers, and radar warning devices. They act as enemy planes during practice dogfights and evasion drills, dodging defensive maneuvers.

Figure 52: Flight International Learjet with threat and jamming pods (Hamel and Park, 2022, 166)

Other training flights feature Learjets towing targets for surface-to-air or air-to-air gunnery practice. Hitting an unmanned drone towed behind a Learjet at combat speeds helps hone critical targeting skills.

Figure :53 Finnish Air Force UC-35A with sea search Radar after a target towing flight (Credit Kai Krause) (Hamel and Park, 2022, 173)

Learjets also play the enemy role when testing ship defense systems. Approaching naval vessels, they simulate anti-ship missiles with special transmitters so sailors can practice intercepting the threats.

Figure 54: Tactical training Learjets operated by the German Gesellschaft für Flugzieldarstellung (GFD), an Aerial Target Simulation Organization (Hamel and Park, 2022, 162)

These aggressive Learjet "aggressors" are vital to ensuring military readiness. They provide realistic battle simulations across air, ground, and sea forces using the latest electronic warfare and targeting technologies.

Figure 55: Intercept exercise with Eurofighters (Hamel and Park, 2022, 163)

Figure 56: GFD Learjets 35A with Laser designators (Hamel and Park, 2022, 163)

Figure 57: Japan Self Defense Force U-36A with a variety of mission equipment (Hamel and Park, 2022, 171)

Figure 58: Saab's Special Flight Operations (SFO) Learjets provide extensive training capabilities, including electronic warfare, target towing, and adversary flight profiles. The Learjet 35As are equipped to tow radar-augmented targets for air-to-air and surface-to-air missile practice. SFO supports military training for several nations' forces. With specialized electronic and towing systems, these versatile jets serve as adversary aircraft to test skills and readiness. (Hamel and Park, 2022, 174)

Learjets keep pilots and defenses sharp and ready thanks to their versatility as adaptable training platforms.

But they also serve as eyes for the military.

With their spy-plane capabilities, Learjets conduct surveillance and gather intelligence from the air. How have specialized Learjets been outfitted for observation missions?

Eyes in the Sky

Learjets serve as high-flying observation platforms for reconnaissance (exploring to gather information) and surveillance missions. Their capabilities suit them well for information-gathering flights.

Outfitted with cameras, sensors, and mapping equipment, Learjets undertake photographic surveys and geological studies. Detailed aerial imaging provides valuable intelligence otherwise hard to obtain from the ground.

Figure 59: Phoenix Air Learjet 35A "Gray Bird." Phoenix Air uses specially equipped Learjets for military training worldwide. The "Gray Birds" have systems to jam radar and communications. They can carry pods and tow targets. This lets them simulate enemy threats against planes, ships, and ground bases (Credit Achim Stemmer) (Hamel and Park, 2022, 167)

NASA mounted telescopes in one Learjet used for infrared astronomy research. Others carried equipment to digitally image landscapes and urban infrastructure after disasters.

Police also use Learjets to help find criminals and collect evidence. Their speed allows them to rapidly respond and quietly observe situations as they develop.

But Learjets aren't just flying cameras. Custom instruments allow them to sniff out data on airborne chemicals, pollution, and particles during research flights.

All of these observation and data gathering missions rely on the Learjet's reliability, range, and ability to manage specialized payloads. Learjets serve as versatile sensor platforms gathering hard-to-reach information high above the clouds.

While Learjets excelled as workhorses on serious missions, some pilots sought to exploit the jets' agile abilities in another way: airshows.

Legendary pilots performed heart-stopping aerobatics and stunts in Learjets, revealing their incredible agility. What daring feats have bold aviators accomplished with Learjets?

Unleashing Learjet's Wild Side

While designed as business jets, Learjets retain some of the speed and agility of their P-16 fighter jet ancestors (as discussed in Chapter 2). This made them prime targets for daring pilots to push to new limits.

Figure 60: Clay Lacy's Learjet Model 35A (Credit Tomás Del Coro) (Hamel and Park, 2022, 189)

Legendary pilot Clay Lacy used his skill and precision to perform jaw-dropping aerobatic shows in Learjets. During airshows, he would put the planes through twists, turns, rolls, and dives that seemed to defy gravity. Watching the shiny silver Learjet zoom upside down just hundreds of feet above the ground made the crowds gaze skyward in disbelief.

Figure 61: Avstar Lear Jet 24A with new registration-code N3137 (Credit Nick Dean) (Hamel and Park, 2022, 192)

Lacy's airshow routines demonstrated the incredible maneuverability hidden within the Learjet's typically businesslike exterior. His breathless aerial stunts revealed the adventurous spirit at the heart of Learjet aviation.

Figure 62: Younkin's Lear Jet 23-009 during aerobatics (Hamel and Park, 2022, 194)

Other bold pilots followed Lacy's lead. Aviator Bobby Younkin modified a Learjet 23 just for aerobatics. This custom "Learobatic" jet could turn, tumble, and streak across the sky with ease thanks to Younkin's skill. His daring performance showed crowds just how agile the Learjet could be.

Figure 63: Younkin barrel roll with Learjet 23-009 (Credit Adam Wright) (Hamel and Park, 2022, 195)

While not intended for stunts, Learjets have proven themselves more than capable in the hands of audacious pilots. Their sleek frames conceal an incredible capacity for speed, agility and thrills when these jets take to the skies.

Figure 64: FlyersTeam Learjet 24 in 2006 colors (Credit Ivan Alejandro Hernandez Velasco) (Hamel and Park, 2022, 196)

Figure 65: FlyersTeam rolling Learjet 24 in 2011 colors (Credit Ivan Hernandez) (Hamel and Park, 2022, 197)

Conclusion

Figure 66: Learjet model 55C with delta fins (55C-136, D-CFAZ) (Hamel and Park, 2022, 113)

The Learjet shined in diverse special missions. Many of the airplane fight scenes you see in movies were filmed using cameras on Learjet planes owned by Clay Lacy. In addition to being used for filming, Learjets serve a wide range of functions.

For example, they have been used all around the world to quickly transport sick or injured people to hospitals. Furthermore, militaries have used Learjets for pilot training, mapping from high altitudes, and spying missions. One specific movie that showcases Learjet camera footage is the exciting fighter jet scenes in *Top Gun*.

Daring pilots even performed heart-stopping aerobatics at airshows in the agile jets.

Though created for business travel, the Learjet exceeded expectations. It proved a platform for scientific research, adversary training, aerial photography, and other innovative duties. The jet's reliability and performance allowed it to take on challenges beyond normal civilian flights.

The Learjet rose to the occasion thanks to the unsung engineers whose creativity made it fly. Their inspiring story of overcoming obstacles to innovate is a tale of imagination and perseverance. It's the story of how a small team achieved the extraordinary through dedication to a dream.

Next, we'll go behind the scenes for a look at some more off-beat uses.

Key Takeaways

- ✓ **Movie Stars of the Sky**: Learjets have been used in Hollywood movies to film amazing action scenes in the air.

- ✓ **Innovative Cameras**: Famous pilot Clay Lacy invented a new camera system called Astrovision. This camera could rotate 360 degrees and capture every angle during a flight!

- ✓ **Capturing Fast Jets**: Learjets were used to photograph really fast planes. But one time, a tragic accident happened when one jet crashed into another during a photoshoot.

- ✓ **NASA's Fast Labs**: NASA used Learjets for science experiments. They tested new airplane technologies and even studied space!

- ✓ **Training Military Pilots**: Learjets play the role of "enemy planes" to train military pilots. They are equipped with special tools to make the training as real as possible.

- ✓ **Target Practice**: Learjets also help in practicing shooting targets in the air. They tow targets behind them that military pilots try to hit.

- ✓ **Eyes in the Sky**: Some Learjets are used to gather important information from high up in the sky. They have cameras and sensors to take detailed pictures of land and even to help police.

- ✓ **Aerobatics and Stunts**: Some daring pilots use Learjets to perform breathtaking stunts in airshows, showing how agile and versatile these jets are.

- ✓ **Versatile Missions**: Whether it's for movies, science, military training, or stunts, Learjets can do a lot of different things really well.

Chapter 6
Learjets Take Hollywood

Introduction

Get your popcorn ready as we soar into how Learjets lit up the silver screen! This chapter reveals how Hollywood blockbusters used Learjets to capture thrilling aerial action scenes.

You'll read how celebrity pilots like Clay Lacy spread Learjet's fame. Read on for legendary tales of Learjets and the stars who flew them!

Figure 67: Golfer Legend Arnold Palmer (Hamel and Park, 2022, 305)

Lacy's Famous Clientele

Bill Lear suggested to Clay Lacy that he give celebrities Learjet rides. So Clay flew Hollywood stars in Learjets through his charter company. He also flew camera planes for movies as seen in Chapter 5. He transported stars to Las Vegas and other places. Famous people like Frank Sinatra, Dean Martin, and President Ronald Reagan flew with Clay Lacy Aviation.

Figure 68: Clay Lacy's Learjet 24A prepared for the Palm Springs Air Museum (Hamel and Park, 2022, 191)

Lacy's flights weren't always just about getting from point A to point B. He once urgently flew a beaten-up Sinatra and Martin away on his Learjet after a wild party. For decades, his charter service flew Hollywood's elite. This helped make the Learjet super popular.

Figure 69: Clay Lacy (left) and Bill Lear (Hamel and Park, 2022, 190)

Fueling the Stars: Learjets Take Off in Hollywood

Sinatra's Jet Set Rides

After experiencing the Learjet with Clay Lacy, Frank Sinatra bought one of the first Learjets in 1965 - a Learjet 24A. This helped make Learjets super popular with other Hollywood stars too. Sinatra loved his Learjet's speed. He could zip all over the country for his shows and parties.

Figure 70: Frank Sinatra's Learjet 24A (N175FS) purchased in 1966 (Hamel and Park, 2022, 256)

Sinatra named his Learjet "Christina II" after his daughter.

But it wasn't just the speed that Sinatra adored about his Learjet. He flew his Rat Pack buddies like Dean Martin and Sammy Davis Jr. on adventures in it. One time he used it to fly them to march with Martin Luther King.

Figure 71: Sinatra's Learjet ferried Dean Martin (left) and other members of the Rat Pack (Hamel and Park, 2022, 256)

Sinatra's Learjet was more than just a fast mode of transport. It also played a role in his personal milestones.

When Sinatra proposed to Mia Farrow, he hid the ring on his Learjet. Later they flew in it to France for their honeymoon. Sinatra even loaned his Learjet to Elvis Presley. Elvis used it to quickly marry Priscilla in Las Vegas.

Sinatra wasn't the only big name connected with Learjets; other celebrities had their own tales to share.

Figure 72: Sinatra's Learjet N175FS enabled a speedy Wedding in Las Vegas (Credit Bettmann Archive) (Hamel and Park, 2022, 258)

Figure 73: Elvis and Priscilla in Sinatra's Learjet (Credit Bettmann Archive) (Hamel and Park, 2022, 258)

While Lacy and Sinatra made headlines, Learjets also had a part to play in more somber moments in Hollywood history. For example, the story of Howard Hughes shows that Learjets were also used in less happy but deeply important ways.

Howard Hughes' Last Flight

In 1976, a Learjet 24 flew former Hollywood producer and billionaire Howard Hughes on a final secret trip. Once a popular Hollywood star and famous pilot, now Hughes looked tired and thin in his pajamas on the plane.

Figure 74: Howard Hughes was an aviator, engineer, industrialist, and Hollywood producer. He set air speed records in aircraft he designed, including the Hughes H-1 Racer seen here (Hamel and Park, 2022, 254)

Hughes had directed and produced major films like *Hell's Angels*. In addition, he made a fortune in business ventures. He lived an eccentric lifestyle later in life, but his early aviation innovations left a legacy that long outlived him.

The pilots remembered Hughes as weak and pale, nothing like his glory days. Though the public knew little of his last years as a loner, the Learjet gave Hughes one last smooth private flight before he passed away.

He died at age 70 during the flight to Houston.

The role of Learjets in Hollywood wasn't limited to glamorous moments. They were part of some of life's most final journeys as well.

Get Ready for a Detour

In our next chapter, we'll see what happened when Learjets veered off course into uncharted territory. Without Hollywood glamour lighting the way, some Learjets found themselves on the wrong side of the law.

From drug smuggling to faking pilot credentials, we'll uncover wild stories of Learjets gone rogue.

Key Takeaways

- ✓ Learjets played a starring role in Hollywood films, carrying innovative cameras to capture thrilling aerial action scenes.

- ✓ Clay Lacy became famous for using Learjets to transport celebrities and film movie magic. This increased Learjet's popularity.

- ✓ Frank Sinatra purchased one of the first Learjets in 1965, naming it after his daughter. His Learjet ferried Rat Pack members on adventures.

- ✓ Sinatra loaned his Learjet to Elvis Presley to enable his quick Las Vegas wedding to Priscilla.

- ✓ Billionaire Howard Hughes took a final flight in a Learjet before passing away in 1976 at age 70.

Chapter 7
Learjets Off the Beaten Path

Introduction

This chapter reveals the wild adventures and shady schemes that gave Learjets surprising second lives. You'll read how smugglers like Pablo Escobar used Learjets to secretly shuttle illicit cargo.

These speedy jets even masqueraded as luxury limos, eccentric hybrid planes, and police spy planes!

Not all Learjets followed a straight and narrow path. From leisurely luxury to outright lawbreaking, this chapter explores tales of Learjets that strayed from their intended mission.

Prepare for surprising stories that show Learjets' versatility to adapt when taken in unexpected directions!

Learjets' Shady Side

Escobar's Cash Learjet

Pablo Escobar was a famous drug smuggler in Colombia. He made billions from selling cocaine. Escobar had to get the piles of cash he earned back to Colombia somehow.

Figure 75: Escobar and his wife leaving his cash-money carrying Learjet (Credit cash. imgur.com) (Hamel and Park, 2022, 244)

So in the 1980s, he bought a Learjet! Escobar used the speedy Learjet to zip back and forth hauling suitcases stuffed with cash. The Learjet could fly a maximum of six passengers at speeds under the sound barrier, up to 530 mph, and could cruise at over 480 mph.

The cops couldn't catch his cash-packed Learjet as it flew over the ocean to Colombia. For Escobar, the Learjet was just a way to move his money.

Busted for Marijuana

Recently, police busted a Learjet carrying a huge illegal shipment. The Learjet was flying from California to Florida. But it stopped to refuel in Louisiana in 2021.

Figure 76: Confiscated Learjet 55 packed with pot intercepted on June 23, 2021 after it stopped at the Hammond Northshore Regional Airport (Hamel and Park, 2022, 242)

The police searched the plane and found duffel bags stuffed with marijuana. There was over $1 million worth of illegal weed on board! The cops arrested the passenger and took the pot-filled Learjet away.

More Shady Schemes

In 1990, actor Emilio Estevez used a Learjet to fly to Bolivia. But when he left, the Bolivian police kept his Learjet! They said it had traces of cocaine inside. Estevez said he didn't know about any drugs. But the cops still kept his plane.

Another time, police found a whole 300 kg of cocaine hidden on a Learjet. The crooks used the speedy Learjet to secretly fly drugs around. But the police eventually caught them and took their drug smuggling jet away.

The Tale of Ali Sabouri Haghighi

In 2012, a Learjet 24D crashed in Denmark. Inside the wreck, police found a fake pilot license! An Iranian man named Nader Ali Sabouri Haghighi had crashed the plane. He had a criminal record for plane theft and fraud.

Haghighi used the Learjet 24D for 2 years to illegally fly passengers in Europe. He pretended to be a pilot named Daniel George. When the Learjet crashed, his secret came out. The police arrested Haghighi for his shady flights with the Learjet.

Figure 77: Accident Learjet 24D crashed in a field of sweet corn (Credit AIB DK) (Hamel and Park, 2022, 247)

After jail, Haghighi got out and kept lying. In 2014, he tricked the U.S. Federal Aviation Administration into giving him a new pilot's license in a different name. Haghighi even used the fake license to get a job flying airliners in Indonesia!

Figure 78: Secured Learjet 35 (N31DP) and pilot Haghighi with two guards (Hamel and Park, 2022, 249)

Finally in 2015, Haghighi was caught again and put in jail for 2 years for his continued illegal flying. However, this sneaky pilot kept causing trouble with his criminal Learjet schemes.

Odd Jobs: Learjets' Unconventional Missions

Limos, Hybrids and Spin-Offs

Folks found creative new uses for old Learjets. One became a Learjet limo! With leather seats and disco lights, it was a party on wheels.

Figure 79: Learmousine (Credit Mecum Auctions) (Hamel and Park, 2022, 264)

Engineers even built weird hybrid planes using Learjet parts. One was a Lear-stang - part Learjet, part Mustang fighter. It had a Learjet wing and Mustang body. Sadly, the Lear-stang crashed at an airshow in 1999. But it showed Learjets could do unexpected things!

Figure 80: P-51 Miss Ashley II with Learjet wing and tail and contraprpops (Credit D. Thirot) (Hamel and Park, 2022, 263)

Conclusion

While Learjet made its jets for luxury and business, folks found other fun uses too! Learjets strayed down colorful new paths as fancy limos, weird hybrid planes, drug mules, and secret police rides.

Whether used for right or wrong, Learjets showed they could take on any job. Their speed and power worked great for real executives or sneaky smugglers.

In the end, Learjets could do it all, expected or surprising. This flexibility made them a jack-of-all-trades tough enough to go almost anywhere. Wherever Learjets flew, they made an adventurous legacy.

Now get ready for more tales from the Learjet's past. The next chapter will take us behind the scenes with the engineers who made these machines magically fly. With ingenuity and teamwork, they overcame challenges to create the legendary Learjet. Get ready for inspiring lost tales of how ordinary folks worked wonders and gave us wings!

Key Takeaways

- ✓ Pablo Escobar infamously used Learjets to secretly shuttle cash from his illegal drug operations. The jets' speed and range enabled covert transportation.

- ✓ Learjets have been involved in other criminal activities like marijuana smuggling and cocaine trafficking thanks to their capabilities.

- ✓ Some Learjets have taken on unconventional second lives as luxury limos, eccentric hybrid planes, and police planes.

- ✓ Not all Learjets have followed their intended purpose, sometimes straying into lawbreaking or curiosity-sparking new uses.

- ✓ The stories show the versatility of Learjets to adapt beyond business travel into surprising new roles, whether legal or illegal.

- ✓ While the original Learjet opened new possibilities for executive travel, some have found possibilities on the fringes of the law.

- ✓ The tales reveal fascinating, though not always glamorous, lives of certain Learjets beyond their expected realm of corporate jets.

Chapter 8
Learjet: The Lost Tales

Introduction

Prepare for takeoff into the untold stories of Learjet! Discover the lost tales of these aviation innovators who dreamed up planes faster than any jet before.

Forget what you thought you knew about the early days of flight. These rarely-shared stories — adapted from tales told by former Learjet family members — will surprise you. Stories packed with drama, near-disasters, and mysteries from those who lived them. Lessons we didn't want lost in time.

Have you ever sat in class and asked how you would ever use what you learned? Well, these Learjet legends are here to show you how their own experiences, talents, and knowledge shaped innovation. From art class to tackling bad weather, you'll see how knowledge paid off in unexpected ways.

There are reasons for every class, even if you can't see them yet. The tales of Learjet's past have a lesson for all of us – you never know when something you learn will take flight to change the future.

So get ready for a record-breaking ride into the untold past of aviation's envelope-pushing innovators as we rediscover Learjet's lost tales.

Mike Abla's Shakey Parlor

Mike Abla, former director of technical engineering at Learjet

This is the story of how art class paid off for one enterprising engineer.

Mike Abla was a super smart engineer who worked at Learjet long ago. He wanted to build safest planes around. But to do it, he needed ground vibration test (GVT) equipment (a plane shaking machine).

This amazing machine could shake a whole plane to see where it might vibrate too much. Mike could use it to make changes to keep important parts from breaking and the wings from falling off.

The GVT equipment Mike wanted would cost a ton of money. Two million dollars! When Mike asked his bosses for the cash, they said "No way!" They thought it was too expensive.

Mike was bummed. But he wasn't ready to give up on his dream. He decided to show the bosses why he really needed that equipment.

Mike imagined his co-workers shaking the planes themselves, since they didn't have the equipment. The bosses doing silly things like using a jump rope over the wings! He drew a funny cartoon of his idea and labeled it "Abla's Shakey Parlor" - a place where people pretended to be a shake machine.

When the big boss saw Mike's hilarious cartoon, he finally got the point. Building planes without the right equipment was ridiculous! The boss changed his mind and agreed to buy Mike's multi-million dollar GVT equipment.

With his cool new equipment, Mike made Learjet planes safer than ever. His innovation paid off big time! Mike proved that with creativity and persistence, an engineer can shake up even the highest-flying company.

Figure 81: Mike imagined a "Shakey Parlor" where employees would have to manually shake the planes, like the Learjet president jumping rope on the wing in his cartoon. The funny nickname helped convince the bosses to invest in Mike's vibration testing equipment! (Hamel and Park, 2022, 273)

The Mystery of the Barrel Roll Buzz

Mike Abla, former director of technical engineering at Learjet

Learjets were crashing, and no one knew why. Each crash looked the same. The planes were smashed in tiny craters. Their elevators - the flaps on the tail that control up and down - were torn off and missing.

Figure 82: Elevator failures were almost identical. The elevator was broken inboard of the elevator horn (Hamel and Park, 2022, 270)

It was a real mystery. Learjet's top engineers, Mike Abla and Pete Reynolds, wanted to crack the case. They listened closely to recordings from the broken planes. Over and over they heard a weird buzzing noise.

"That's it!" realized Mike. He knew that sound from flying jets insanely fast. It was aileron buzz, caused by the plane's flaps on the wings.

Mike and Pete took test flights beyond the speed limit to recreate the buzz. Their recordings matched the crashing planes. Now they had to figure out why the buzz led to disaster.

The answer came when a pilot landed fast with a busted elevator. He confessed to disabling the speed warning and flying too fast - faster than the plane was built to fly.

Mike solved the puzzle. No more speeding and no more mysterious crashes.

It was years later that the co-pilot admitted they were doing barrel rolls at night when the incident happened.

Just Reverse the Hook

Stan Blankenship, Chief of Sales Engineering

All seemed well as the sleek Learjet descended toward Wichita. But something was wrong. The nose landing gear refused to lower into place.

On the ground, emergency crews sprang into action. Fire trucks raced onto the runway, laying down strips of foam for the troubled plane to attempt an emergency landing.

Guided by the steady pilots, the Learjet's nose skidded down the foamed runway and came to rest with little damage. After the dramatic landing, Learjet engineers inspected the plane. They spotted the fix in minutes.

A hook had been designed to hold the gear up if hydraulics failed mid-flight. But in this case it prevented the gear from coming down due to low pressure in the struts. So the engineers came up with a clever fix for future flights. They simply turned the troublesome hook around.

This easy innovation prevented future landing scares.

Figure 83: Belly-landed Learjet, now all fixed-up (Hamel and Park, 2022, 323)

The Test Flight on the Edge of Disaster

Learjet engineers had a dangerous job testing planes' safety systems. They had to make sure things worked in an emergency.

One day, experienced pilots Jack and Hank strapped into the cockpit. Today they would test the emergency oxygen masks.

At high altitudes, pilots can't breathe without oxygen pumped into their masks. Learjet wanted to know what would happen if the system failed mid-flight.

Jack and Hank agreed Hank would turn on the oxygen to simulate failure. They'd put on emergency masks and Hank would dive the plane fast to lower altitudes.

But when Hank cut the oxygen, both pilots blacked out! The plane was now screaming through the sky with two unconscious pilots.

Luckily, Hank's younger copilot came to first. He took the controls and pulled out of the dive just in time. By risking their lives testing safety systems, these brave pilots had almost met disaster.

A Cat Steers the Pilot

Back in the early days of aviation, pilots flew with only basic instruments in the cockpit. They had to rely on their senses and intuition to fly safely.

One of those daring pilots was Austin Goodwin, known as "Goody." Goody was a big guy who looked like a farmer. He often wore overalls, work boots, and flannel shirts. He was known for his friendly personality. Goody was comfortable flying by feel and intuition even in challenging conditions.

After World War II, Goody flew all kinds of aircraft using just a few key gauges. He worked as an instructor pilot for the new Learjet company in the 1960s.

Goody had an unusual copilot - his pet cat! That's right, his cat "Kittyhawk" (not its real name) flew everywhere with him. Kittyhawk would curl up on Goody's lap for their adventures in the skies.

On night flights through cloudy skies, Kittyhawk helped steer the plane. If Goody slipped up and didn't make a "coordinated turn", Kittyhawk dug her claws into his leg as a warning. Those sharp claws got Goody back on track!

What's a coordinated turn? It's when a pilot uses just the right amount of rudder with aileron to keep the plane turning smooth and straight. Uncoordinated turns feel like a car turning a corner. The more uncoordinated, the stronger the passengers are pushed to the side.

Can you imagine flying an airplane while a cat sits on your lap? Goody and Kittyhawk made an unlikely team. But in those early flying days, pilots learned to trust their instincts - and sometimes their furry friends too! Kittyhawk's claws kept them flying straight and true through dark skies when only a few gauges lit up the cockpit.

So next time you get on a plane, remember pilots long ago like Goody and his cat, flying the friendly skies together!

Flying Cowboys Meet Their Doom

In the wild early days of Learjet, danger lurked for operators who ignored the rules. Some were real cowboy flyboys, taking huge risks just for adventure.

These flying outlaws installed secret "go-fast" switches in the cockpit to disable warnings. With a flip of the switch, they could blast past safe speeds.

But the danger was real. Their need for speed turned deadly when Learjets started crashing mysteriously. By the time investigators discovered the go-fast outlaw switches, it was too late for some.

These flyboys didn't understand the real risk - "Mach tuck." Past the speed limit, deadly forces could seize control. As the Mach number, their speed vs sound, increased uncontrollably, the nose pitched down and the plane violently spiraled until it broke apart.

Wrecked planes told the tragic story: wings sheared off, tailfins torn away in mid-air. Going too fast had destroyed them. When officials grounded the renegade Learjets, the fun and games stopped.

The outlaws learned safety comes first, no matter how tempting the call of high-speed adventures. Rules exist for a reason, even in the wild blue yonder.

The Miracle on Lake Michigan

It was a clear but chilling morning on March 19, 1966 when the Learjet 23 took off from Chicago's Meigs Field. The sleek silver jet climbed smoothly into the blue sky, unaware of the danger that lay ahead.

Just six months earlier, Learjet Corporation had been reborn as Lear Jet Industries. The young company's future looked bright, led by aviation pioneer Bill Lear. His Learjet 23 was set to revolutionize business travel with its speed and efficiency.

But on this fateful flight, the Learjet was piloted by a crew from Mutual Benefit Health and Accident. Their flight plan took them north over Lake Michigan. As they reached cruising altitude, a thin layer of ice began forming on the wings. The pilots turned on the jet's anti-icing system. But it was too late.

The ice built up rapidly, disrupting the smooth flow of air into the engines. Flames suddenly burst from the jets with a roar. In an instant, both engines flamed out. The Learjet went silent, the only sound the rush of freezing wind.

The stunned pilots had no choice. Gently, they guided the powerless jet into a glide. Below, the gloomy gray waters of Lake Michigan awaited. The copilot sent out a Mayday call as the captain aimed for the white-capped swells.

The Learjet splashed down softly, slowing to a stop as its nose bobbed in the waves. To the delight of an eyewitness, it was "the most beautiful amphibious landing he had ever seen."

By a miracle, the jet remained afloat for 24 hours, battered but intact. The next morning, a recovery crew arrived to tow the Learjet to a boat ramp.

Figure 84: Lear Jet 23-061 (N316M) being towed (Credit Mike Abla) (Hamel and Park, 2022, 65)

As a result, the experience of this event was used by Learjet and extrapolated to other Learjet models for future certifications for compliance with the FAR ditching requirements. Thanks to the skill and poise of her pilots, the lost Learjet survived to fly again another day.

Her ditching stands as a shining moment in Learjet's early history – a tale of innovation and grace under pressure.

Figure 85: Lear Jet 23 floating at a boat ramp (Credit Mike Abla) (Hamel and Park, 2022, 65)

Figure 86: Lear Jet 23 being lifted (Credit Mike Abla) (Hamel and Park, 2022, 66)

The Chicken Chucking Blow-Up

From "Fly Fast-- Sin Boldly"

Boom! The chicken cannon fires. A four-pound bird rockets toward the new Learjet windshield at 250 knots. Wham! It pierces the plexiglass like a missile. Engineers freeze in shock as the chicken blasts out the back screaming for the hills.

Epic fail! Bill Lear thought Lexan was unbreakable. He'd jumped on it. He dropped cannon balls on it. He even made an employee shoot it with a pistol. But a four-pound chicken demolished Lear's windshield.

"I wonder if the FAA would accept a 250-pound chicken at just 4 knots?" Lear wondered. Then the light bulb went on! The problem was momentum.

Mass and high velocity create strong momentum, like a bowling ball hitting pins. The chicken's momentum hit one spot too hard.

But half the mass would create half the momentum. So Lear's team installed a blade on the center windshield frame to slice the chickens in two. Now the impact hits two spots. The next test was a smashing success! Each Lexan pane only had to stop half a chicken.

Figure 87: Bird Gun pointing to the Lear Jet wind shield (Credit Dick Kovich) (Hamel and Park, 2022, 327)

Learjet kept improving designs over the years. Today computers test virtual windshields. But it all started with the tale of a high-flying hen - and using failure to launch innovation.

Boom! What will you learn from failures on your way to success? Each flop brings you one flap closer to flight!

Experience is the Best Teacher

Stan Blankenship, Chief of Sales Engineering

Bill Lear learned an important lesson from his own flying experience. Through years in the skies, he knew the hazards pilots faced in marginal weather when they had to rely on single instruments with no backup.

So when Bill discovered the Navy's planes had angle-of-attack indicators, he understood their value right away.

This gauge could help prevent deadly stalls and warn when wings were too steep.

Unlike "fair weather pilots", Bill grasped the true benefit of this instrument. His first-hand experience made him eager to provide the same protection for Learjet pilots.

The same was true when Bill added backup artificial horizons. He knew from flying blind himself how precious these tools were for staying level if primary gauges failed.

What Bill learned through hard-won sky wisdom, he passed on by equipping Learjets with redundancy. Bill's experience recognized the need for safety and reliability.

No textbook could teach what Bill had lived. By listening to the lessons of his own journeys, Bill gave Learjet pilots the instruments today's pilots now take for granted.

The sky had molded Bill into a thoughtful designer who helped generations yet to come.

Key Takeaways

- ✓ These stories provide insights into Learjet's untold history and early innovations. They capture drama and mysteries almost lost to time.

- ✓ The tales showcase how diverse experiences and skills of Learjet's creators contributed to their creativity and problem-solving.

- ✓ Mike Abla's art talents helped convince bosses to fund needed equipment through an imaginative cartoon. This shows creativity's role.

- ✓ Bill Lear's flying background made him recognize the value of safety instruments that pilots now take for granted. Real-world experience shaped innovations.

- ✓ The stories aim to show students that the relevance of what they learn isn't always obvious immediately. Knowledge can pay off unexpectedly.

- ✓ The pilots' poise during emergencies exemplifies grace under pressure and quick thinking. Their skill saved lives and planes.

- ✓ Rules and safety procedures exist for good reason, as some pilots unfortunately learned the hard way testing limits.

- ✓ Persistence and determination to try new things were key to Learjet's boundary-pushing advances that made it an aviation legend.

- ✓ Capturing these tales helps preserve insight into Learjet's history and transmit innovative lessons to future generations.

Chapter 9
Final Frontiers

Innovative Upgrades: 70 and 75 Models Soar to New Heights

In 2013, Learjet took its Models 40 and 45 to the next level by introducing the advanced 70 and 75. These updated jets built upon the success of their predecessors and offered significant enhancements.

Figure 88: Learjet 70 (Credit Bombardier) (Hamel and Park, 2022, 130)

The 70 and 75 models featured turbocharged engines that delivered extraordinary performance. Their aerodynamically designed winglets made them more efficient, enabling higher speeds and increased fuel efficiency.

Equipped with powerful Honeywell engines, these jets had the ability to climb faster and carry heavier loads.

The interior design prioritized passenger comfort with spacious cabins, featuring flat floors and extra headroom. The 70 model accommodated seven passengers, while the 75 model had room for nine.

Figure 89: Learjet 75 Liberty (Credit Aerokurier) (Hamel and Park, 2022, 130)

For those seeking a more luxurious flying experience, the 75LXi model offered premium features, including high-quality leather seats and finely crafted metal accents. On the other hand, the 75 Liberty provided a simpler, more affordable option without sacrificing comfort or safety.

Figure 90: Learjet 75LXi luxorious interior with ergonomic leather seats (Hamel and Park, 2022, 131)

Overall, the 70 and 75 models demonstrated that Learjet continued to be an innovator in the aviation industry, offering state-of-the-art technology for an enhanced travel experience.

Lear's Winged Wonders - The Learstar and Lear Fan

Bill Lear was bursting with radical ideas for wild new aircraft. Two of his inventive designs became the Learstar and Lear Fan.

The Learstar 600 started as Lear's plan for a twin-engine business jet. Canadair/Bombardier bought the rights and turned it into their tubular Challenger 600 series - a chart-topping hit!

The Lear Fan was Lear's vision for a smooth composite airplane with push-propellers in the back. Lear's vision for this novel airplane could not be completed during his lifetime.

Figure 91: The Lear Fan (Hamel and Park, 2022, 339)

Even though the Lear Fan never fully took off, it showed Lear could dream bigger than big. His head was always buzzing with new inventions to pioneer the skies!

These imaginative planes proved Bill Lear's creativity knew no limits. He kept reaching for the stars, inspiring others to aim high and try the unconventional.

Legends Falter - The Ill-Fated 85

In 2007, Learjet embarked on their most ambitious jet ever - the advanced Model 85. This large jet was designed to push limits.

The Model 85 was constructed with cutting-edge composite materials. It had an extended fuselage and enlarged wings for extraordinary range capabilities.

Figure 92: Learjet 85 first flight was on April 7, 2014 (Hamel and Park, 2022, 133)

This expansive jet aimed to carry more passengers at faster speeds and increased distances than prior Learjets. It was an imposing endeavor.

However, developing the formidable Model 85 proved extremely challenging. Numerous issues caused extensive delays and expenses far beyond projections.

After its promising first test flight on April 9, 2014, the imposing Model 85 was cancelled.

The Model 85 did not make it to market, but it showed Learjet's spirit of big innovation. It aimed high and imagined future possibilities. Though the Model 85 never flew for customers, its huge goals reflected Learjet's drive to always move forward.

But those big plans would soon fall apart.

The Model 85 represented Learjet's creativity and determination to keep making business jet travel better no matter the difficulty. Sadly, that would come to an end only 8 years after the 85's first flight.

Learjet's Final Countdown

At first Learjet flew the only business jets in the skies. But over time, competitors squeezed Learjet out.

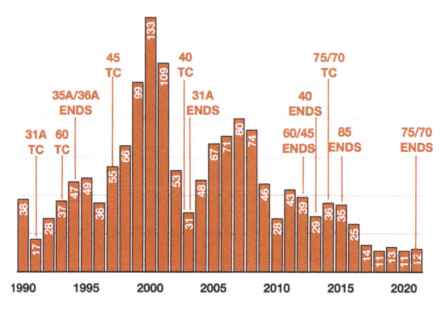

Figure 93: Bombardier Learjets deliveries, TC approvals and production ends (Hamel and Park, 2022, 224) © 2023 The Authors, under exclusive license to Springer Nature Switzerland AG, reproduced with permission

Moving Learjet's production around so much didn't help. Each costly move took time away from building better jets. Learjet faced another disadvantage over other business jet makers. In 1977 the FAA refused to let Learjets fly with only one pilot instead of two.

On March 28, 2022, the last Learjet ever rolled off the line. As employees cheered, an era came to a close after 60 years.

But even as Learjet flights faded into history, their legacy lived on. Learjet pioneered business jets.

Thanks to Learjet's innovation and Bill Lear's vision, luxury jet travel became a reality. Now executives could jet between meetings in comfort.

Though competitors eventually seized the lead, we owe Learjet gratitude for first sparking the business jet revolution. Their spirit of adventure lifted aviation to new heights!

Key Takeaways

- ✓ The cool new 70 and 75 jets kept innovation going with faster engines and cozier cabins.
- ✓ Bill Lear's wild ideas included the Lear Fan, a plane unlike any other! His big dreams had no limits.
- ✓ Even legends sometimes fail – Learjet's mighty Model 85 proved too tough to finish.
- ✓ While Learjet made the first business jets, competitors eventually zoomed by them.
- ✓ We owe Learjet thanks for starting the business jet revolution – they showed luxury flight was possible!
- ✓ Their spirit of adventure lifted aviation to new heights. Learjet was a pioneer that dreamed big.
- ✓ They opened the skies to new ways of flying. Their bold innovation inspired others to imagine.
- ✓ Even though Learjet stopped making jets, their legacy lives on in the history they made.

Chapter 10
Dreaming of Flight

Pre-Flight

You slide into the pilot's seat of a Learjet 23 cockpit. This jet, based on a Swiss P-16 fighter, can carry up to six passengers in luxurious comfort at speeds over 540 miles per hour! It feels like you are an astronaut sitting in a spaceship ready to blast off!

Figure 94: Lear Jet 23 first Flight with Hank Beaird (Credit Neil Corbett) (Hamel and Park, 2022, 345)

The control panel in front of you glows with blinking lights and gauges. There are so many colorful buttons, switches, and dials.

Figure 95: Learjet 60 cockpit design (Credit Carlos Alberto Rubio Herrera) (Hamel and Park, 2022, 117)

Before taking off, you carefully check to make sure everything is working right. You do a walk around the smooth, silvery Learjet 23, testing the flaps and wings to see that they move like they should. When you start the engines, you feel the power rumbling and roaring beneath you.

After taxiing onto the runway, you take a deep breath, excited for what's coming next. Your hand grips the thrust lever, ready to release the massive energy in the engines.

Figure 96: Lear Jet 24 taxiing (Credit Eduardo Capdeville C) (Hamel and Park, 2022, 39)

Cleared for Takeoff

With the engines screaming loudly, you gently pull back on the yoke. The jet blasts forward and leaps into the air! As the ground drops away below you, you look out the big windows at the bright blue sky ahead.

You know this Learjet cockpit is your gateway to speed, height, and adventure - it can climb to 40,000 feet in just seven minutes! Beyond your wildest dreams!

Figure 97: Learjet 35A flying overhead (Credit Andrei Bezmylov) (Hamel and Park, 2022, 67)

As you soar through the clouds in the innovative Learjet 23, it's hard to imagine the long road that made your flight possible. Now let's look back at how one man's ambitious dream first gave wings to the Learjet.

Bill Lear's Big Dream

Imagine Bill Lear, who made your flight possible, dreaming of revolutionizing business travel in the 1950s. The sleek lines of a jet plane fill his mind. He sketches designs, makes plans, and searches for the right team to make it real.

Figure 98: P-16 and Lear Jet 23 at Altenrhein 1965 (Hamel and Park, 2022, 52)

Challenges arise at every turn. Cultural clashes hamper manufacturing overseas. He burns through money faster than it arrives. But Lear pushes forward, undeterred.

His bold imagination envisions an airplane unlike any other - small yet fast, affordable yet luxurious. Driven by curiosity and ambition, he assembles thinkers and tinkerers as eager as him to create something new.

Together, they innovate and experiment. Careful calculations shape graceful wings and a sturdy airframe. Meticulous mechanics install powerful jet engines. Futuristic features emerge - a bullet-nosed cockpit, an aerodynamic T-tail, even a clamshell passenger door.

Figure 99: Bill Lear and his Learjet 24-133 with the innovative clamshell door (Hamel and Park, 2022, 338)

Against the Wind

Setbacks try to stall his momentum. Prototypes crash, investors waffle, regulations require reworking. But giving up is not in Lear's DNA. He finds workarounds, tweaks designs, and presses on.

Touching the Sky

Until finally, his vision takes flight. As the first gleaming Learjet lifts into the sky, Lear watches his sky-high dream become reality. Imagination takes wing on the power of perseverance.

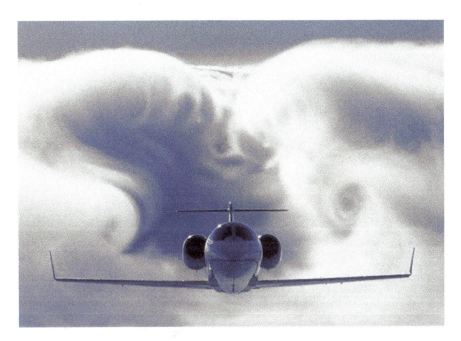

Figure 100: Learjet 60 flies gently bedded between Tip Vortices (Credit Paul Bowen) (Hamel and Park, 2022, 343)

This is the story of every innovator who refuses to be deterred. Of seekers unafraid to fail on the way to success. Of bold minds that envision a future unbounded by today's limits.

Imagine That

What far-fetched idea lives in your imagination? What challenge will you tackle with relentless creativity? How will you turn your dreams to flight?

What will keep you from giving up? Imagine that.

Dreaming of Flight

Figure 101: William Powel Lear (1902-1978) as seen in the Hall of Fame Hallway at the San Diego Air & Space Museum (Hamel and Park, 2022, 342)

Personal Reflections

Are you working on a book report, presentation, or essay? Take a moment to reflect on the broader implications of what you've learned. The story of Learjet isn't just about a single company or even just about aviation; it's a tale of innovation, technology, and the human spirit.

The questions below are designed to help you relate the material in this book to bigger themes and ideas. Feel free to jot down your answers, discuss them with friends or family, or even bring them up in your classroom for a group discussion.

On Innovation

- What was the most surprising innovative leap made by Learjet, in your opinion?

- How do you think the willingness to innovate has helped Learjet remain relevant over the years?

- Can you think of other industries where innovation has played a critical role in their history? How do they compare with Learjet?

On Technology

- How has technology evolved in the aviation industry since Learjet's inception?

- What are the benefits and potential drawbacks of rapid technological advancements in aviation?

- How do you see technology shaping the future of air travel?

On Risk and Reward

- What risks did Learjet or its founders take that you found particularly bold? Were these risks rewarded?

- Do you think innovation is possible without some level of risk? Why or why not?

- Can you name a time when taking a risk led to a reward in your own life?

On Social Impact

- How did Learjet contribute to social or economic changes, locally or globally?

- What lessons can we learn from Learjet about the potential social impacts of business decisions?

On Personal Goals and Ambitions

- Has learning about Learjet inspired you in any way? How?

- What personal goals could be achieved through embracing innovation and risk-taking in your own life?

Feel free to use these questions as a launching pad for your own deeper reflections and discussions. Sometimes, the story of one company's rise and challenges can offer us valuable lessons that apply far beyond the world of aviation.

For more fun ideas, visit:

https://learjethistory.com/ideas.

Thank you for joining us on this journey through the history of Learjet.

References

The chapters in this book were adapted primarily from the extensively researched aviation history text *The Learjet History: Beginnings, Innovations and Utilization* by Peter G. Hamel and Gary D. Park.

The Learjet History served as the foundation and provided in-depth details for the key events, aircraft models, and pioneering figures covered in this book. The authors used and adapted content from their definitive work to make Learjet's fascinating story accessible to younger readers.

While simplified for a 10- to 12-year-old reading level, the accounts in this book aim to accurately represent the technical achievements and captivating history of Learjet aviation. Please refer to the chapters of *The Learjet History* listed below for the original source material and academic references.

The concise references provided connect back to relevant sections of *The Learjet History* for readers interested in further exploring the engineering feats and dramatic tales that shaped the dawn of the business jet age.

References:

Hamel, Peter G, and Park, Gary D. 2022. *The Learjet History*. Springer. Figures reproduced with permission from Springer Nature Switzerland AG. © Springer Nature Switzerland AG

2023. Discover more information about this comprehensive book at https://learjethistory.com.

Chapter 1: Introducing the Learjet Adventure

References Ch 2

Chapter 2: Sky-High Dreams

References Ch 2, 3, 4.1, 4.2

Chapter 3: Pushing the Limits

References Ch 4, 6.2, 6.4, 8.3, 8.5

Chapter 4: Taking Flight to New Heights

References Ch 9.2, 9.3, 11

Chapter 5: Learjet Special Missions

References Ch 9, 12, 13

Lear, William P. 2000. *Fly Fast-- Sin Boldly : Flying, Spying & Surviving* : Lenexa, Ks: Addax Pub. Group.

Chapter 6: Learjets Take Hollywood

References Ch 11, 15.7, 15.9

Chapter 7: Learjets Off the Beaten Path

References Ch 15

Chapter 8: Learjet: The Lost Tales

    References Ch 5.1, 16

Chapter 9: Final Frontiers

    References Ch 8.8, 8.9, 8.10, 14, 18

Chapter 10: Dreaming of Flight

    References Ch 2, 4, 18

About the Authors

Peter Hamel and Gary Park stand alone as the definitive experts on the complete history of Learjet. Their meticulously researched, 456-page book *The Learjet History: Beginnings, Innovations, and Utilization* has been praised as the most comprehensive work ever on this aviation pioneer.

As an aerospace engineer and author, Dr. Hamel brings decades of experience in advanced flight systems. His leadership roles in national aeronautics research make him uniquely qualified to document Learjet's technical feats.

Meanwhile, Gary Park provides insider perspective from over 50 years as an aerospace engineer in Kansas. Having worked on Learjet certification himself, his firsthand knowledge is unparalleled.

Together, their book weaves extensive original interviews, archival material, and technical insights into the definitive Learjet chronicle. Their personal connections in the aviation community gave them exclusive access to details no one else has published.

For young readers, Hamel and Park have distilled their exhaustive research into the most compelling narratives and visuals. There simply is no other source that can match their combination of technical expertise and passion for sharing Learjet's untold stories.

When it comes to Learjet history, nobody does it better.

Relive the Sky-High Dream
Join the Community Rescuing The First Lear Jet Ever Delivered from the Edge of Oblivion

- ✓ **Historic Resurrection:** Witness the restoration of the first delivered Lear Jet, saved from vanishing from history with the help of Mr. Clay Lacy himself.

- ✓ **Community of Enthusiasts:** Connect with dedicated experts united to protect a significant milestone in aviation history.

- ✓ **Live the Legendary Lear Jet:** Revel in the story of the jet that soared into history.

- ✓ **Stay Connected:** Follow the restoration, anticipating the day the Lear Jet graces the heavens once more.

Your Skyward Adventure Awaits!

Engage with a history that propels the future at The Classic Lear Jet Foundation.

Soar into History! Visit https://classiclear.org Today!

 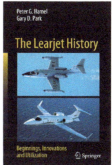

Jim Raisbeck, Dee Howard and *Bill Lear* in 1971 (Hamel and Park, 2022, 98)

Fly Higher: The Comprehensive Chronicle of Learjet Awaits!

Uncover the full spectrum of Learjet's legacy in *The Learjet History*, the definitive extension to your *Age of Learjet* adventure.

- **365 Pages of In-Depth Exploration:** Triple your insight with over 300 additional photos and figures.

- **Extensive Reference Material:** The definitive extension to your *Age of Learjet* adventure.

- **Expanded Anecdotes:** Discover new missions and captivating Learjet tales.

Venture beyond at https://learjethistory.com and start your journey today.

Notes